ADVANCES IN BIOCHEMICAL ENGINEERING

Volume 13

Editors: T. K. Ghose, A. Fiechter, N. Blakebrough

Managing Editor: A. Fiechter

With 134 Figures

Springer-Verlag
Berlin Heidelberg New York 1979

ISBN 3-540-09468-7 Springer-Verlag Berlin Heidelberg New York
ISBN 0-387-09468-7 Springer-Verlag New York Heidelberg Berlin

Library of Congress Catalog Card Number 72-152360
Printed in Germany

Typesetting, printing, and bookbinding: Brühlsche Universitätsdruckerei Lahn-Gießen.
2152/3140-543210

Editors

Managing Editor

Editorial Board

Contents

Application of Microcomputers in the Study of Microbial Processes 1
W. Hampel, Vienna (Austria)

Dissolved Oxygen Electrodes 35
Y. H. Lee, Philadelphia, Pennsylvania (USA)
G. T. Tsao, West Lafayette, Indiana (USA)

Power Consumption in Aerated Stirred Tank Reactor Systems 87
H. Brauer, Berlin (Germany)

Loop Reactors 121
H. Blenke, Stuttgart (Germany)

Application of Microcomputers in the Study of Microbial Processes

Werner A. Hampel
Institute of Biochemical Technology and Microbiology
University of Technology Vienna
A-1060 Wien, Austria

1 Introduction 2
2 Application of Computers to Analyze, Optimize, and Control Microbial Processes 2
2.1 Data Acquisition 2
2.2 Data Reduction and Analysis 3
2.3 Process Control and Optimization 5
3 Microelectronics and Microcomputers 7
3.1 Computer Elements 8
3.2 Minicomputers 9
3.3 Microcomputers 11
3.4 Distributed Computer Network (Multiprocessor Systems) 12
4 Microcomputers Coupled to Bioreactors 12
4.1 Hardware Configuration 13
4.1.1 Microcomputer (Desk-top Calculator) 14
4.1.2 Interface Units for Peripheral Devices 16
4.1.3 Process Periphery 16
4.1.3.1 Data Input Periphery 16
4.1.3.2 Data Output Periphery 19
4.2 Efficacy and Limitations 21
4.3 Cost Considerations 23
5 Broadening Applicability 24
6 Concluding Remarks and Future Tendencies 29
7 Symbols 30
8 References 31

The applications of computers and microcomputers in particular interfaced to bench-top bioreactors are briefly discussed. Compact, quick-access microcomputers (e.g., desk-top calculators) offer the possibility of on-line data acquisition and analysis as well as process control (sequencing, interactive control) with relatively high efficiency at reasonably low costs of installation in laboratory experiments. The configuration of such a computerized system is described in detail along with the capabilities and features of interfacing hardware components. The limits in applicability due to slow operating speeds of fully developed microcomputer systems in particular are pointed out and a survey on investment costs for high-performance, compact desk-top calculators and some peripheral devices is given. Examples of on-line acquisition of several directly accessible environmental process parameters and computations of directly inaccessible state variables are presented in the form of type-writer print-outs. The advantages of on-line experiments for establishing sophisticated control algorithms and for studying the physiological behaviour of the microbial population are demonstrated.

1 Introduction

The advantages and various applications of a system consisting of a computer directly connected to one or several bioreactors have been described in detail in several publications[16, 39, 56] during the last decade and have occasionally been discussed with great vehemence. However, there was no widespread use of such systems which was obviously due to the high installation costs for efficient and reliable computers. It has been possible to increase the utilization of computers in recent years by the development of small, reliable computer systems (minicomputers) thus permitting computer use for pilot plants or even smaller microbial systems at a reasonable price.

The great success achieved with such system configurations has, furthermore, resulted in an increased interest in on-line systems. In the case of bench-top bioreactors the decisive breakthrough was the development of microprocessors and microcomputers respectively. Especially the rapid emergence of compact, quick-access microcomputing systems, e.g., programmable desk-top calculators, capable and flexible enough to meet data processing requirements, made it possible to adapt data acquisition, data reduction and process control even for bench-top bioreactors at reasonable costs. It remains to be investigated as to what extent the microcomputer might replace the minicomputer so far predominantly used in biological growth experiments. In particular, the question posed is which microcomputing system offers optimum possibilities and where the possible limits of the applicability are.

2 Application of Computers to Analyze, Optimize, and Control Microbial Processes

The on-line, real-time utilization of the computer in highly instrumented microbial cultivation systems offers the possibility of

1) collecting large amounts of data derived from various kinds or sensors as part of a data acquisition system,

2) automatically and instantaneously reducing certain data in order to determine the state of the system, as well as of

3) optimizing and controlling the process.

2.1 Data Acquisition

Various methods for determining process parameters based on physical, chemical, and enzymatic analyses have been extensively described[1, 6, 7, 29, 30, 35, 46, 48, 49, 53, 67, 69, 71, 76, 78, 84, 85, 94]. Some of these methods do not fulfill many principal criteria[18] such as reliability, accuracy, reproducibility and prevention of contamination, or they are characterized by a low maximum frequency for handling successively measured values. The latter holds true for several chemical and physicochemical methods (e.g., wet chemical analysis, chromatographic procedures) so that these methods are only applicable to specific, predominantly scientific, problems.

There are only very few adequate systems for continuously and directly *(in situ)* measuring microbial process parameters, i.e., acquiring both physical and chemical factors, such as temperature, pressure, flow rates of gases and liquids, power consumption of the mixing system, pH, rH, dissolved gas concentrations, exhaust air composition.

The application of computers in a data acquisition system does offer some advantages of which the following are pointed out:

Improved accuracy and reliability. Due to statistical methods for obtaining parameter values (e.g., computation of the mean and variance) false and noisy signals can be rejected (digital filtering procedures[8, 45, 47, 77)]). Periodical recalibration automatically performed with several sensor systems may correct drifting.

Increased number of sensing systems. Owing to the possibility of supervising complex measuring devices by the computer, the number of accessible parameters may be considerably increased; e.g., by incorporating batch methods of analysis or by using methods requiring extensive numerical computations for obtaining results.

Cost reduction. Low-cost and simple measuring devices without additional electronic circuits necessary for signal corrections may be used. That is a linearization of signals[6, 30)], compensation of disturbing influences[23, 83)], correction of time lags[21, 69)], etc., are easily performed by the computer in computer-aided data acquisition systems.

Conservation of data. Frequently it is not possible to immediately analyze the information received by the aid of a data acquisition system and thus storage of measured values for reuse is desirable. The length of a single microbial experiment as well as the high frequency of data sometimes necessary results in a large number of values. The computer offers the possibility to store these values on a suitable data base (punched tape, magnetic tape, or disk) after these have been sorted if necessary. These stored data may be used later on, optionally fed to a large mainframe computer, to answer questions concerning the microbial cultivation process (modeling, optimization[11–13, 64, 81, 95)]).

2.2 Data Reduction and Analysis

Several of the monitoring systems are "gateway" sensors, i.e., they open the way through combination with other sensor systems and data for obtaining further information about the microbial cultivation[1, 39, 56)]. Thus, process indicators or directly inaccessible state variables can be obtained which are related to the actual physicochemical, physiological, and biochemical conditions of the culture[57, 60, 61)].

For example, the use of reliable oxygen- and carbon dioxide gas analyzers as well as devices for measuring the gas flow rates and volume of the culture liquid, permit a detailed, real-time study of oxygen uptake and carbon dioxide evolution of microbial cultures. In this respect the bioreactor is comparable to a differential respirometer[1, 23, 60, 61)]. Gas exchange rates are computed according to Eqs. (1) and (2):

$$Q_{O_2} \cdot X = \frac{F_i}{V} \cdot \left(x_i - \frac{w_i \cdot x_o}{1 - (x_o + y_o)} \right) \cdot f, \quad (1)$$

$$Q_{CO_2} \cdot X = \frac{F_i}{V} \cdot \left(\frac{w_i \cdot y_o}{1 - (x_o + y_o)} - y_i \right) \cdot f$$

$$f = \frac{273}{273 + T_i} \cdot \frac{P_i}{1} \cdot \frac{1}{1 + h}, \quad (2)$$

where F_i represents the molar inlet gas flow rate at experimental conditions and w_i, x_i, y_i the mole fractions of inert gas, oxygen, and carbon dioxide, respectively, in inlet air, while w_o, x_o, y_o are the corresponding mole fractions in outlet air. For conversion to standard conditions the absolute pressure (P_i), the temperature (T_i), and the absolute humidity (h) of inlet air must be known. These intermediate data may then be used to compute the volumetric gas transfer rate ($k_L a$) for which additional values of the dissolved gas concentration in the culture liquid are needed. Thus information is provided on existing conditions of mass transfer in the bioreactor at the given operating conditions of agitation and aeration

$$k_L a = \frac{Q_{O_2} X}{(c_L^* - c_L)} \quad (3)$$

On the other hand the oxygen uptake rate and the carbon dioxide evolution rate by the culture can be combined to yield the respiratory quotient (RQ) according to Eq. (4),

$$RQ = Q_{CO_2} / Q_{O_2} \quad (4)$$

thus providing information on the physiological behaviour of the culture. It has been tried several times to establish statistically proven correlations between particular physiological conditions of the microorganism and on-line process information[23, 56, 60]. For instance, the RQ represents a quantitative indicator of ethanol formation and utilization in an aerobic culture of *Saccharomyces cerevisiae.* As it has been described[86], the stages of ethanol formation (RQ > 1.0), oxidative growth (RQ 1.0–0.9), endogenous metabolism (RQ 0.8–0.7) as well as that of ethanol utilization may be distinguished by the corresponding values for the RQ. The coincidence of definite metabolic activities (e.g., nucleic acid synthesis, protein synthesis, ethanol formation) with characteristic changes of the RQ was demonstrated during the transition from lag to the logarithmic phase of growth in the case of a *Candida utilis* culture by means of wet chemical analysis[60].

Heat and material balances not only give insights with respect to the gas exchange of microbial populations[14, 19–21, 26, 40, 41, 86, 87], but by also using additional sensing systems (carbohydrate feed rate, turbidity, amount of neutralizing agents, etc.), information pertaining to the biochemical characteristics of the process can be gained. From data on gas exchange and molasses addition during the cultivation of baker's yeast, online computations of growth rates and different yield coefficients ($Y_{x/s}$, $Y_{o/s}$) were performed using carbon balances[86]. A comparison of the computed cell yields with organic energy yields using different substrates might reflect the mechanism of the breakdown of organic compounds. The correlation between these variables will aid in the elucidation of the main metabolic pathways and reflect the efficiency of the cultivation process[86, 97].

Easily and directly accessible process parameters can be used to estimate directly inaccessible parameters if appropriate models exist (indirect measurement concept[36, 40, 41, 45]). Figure 1 shows the estimation of biomass concentration and specific growth rate from data on exhaust air composition and air flow rate. After computing the amount of assimilated oxygen using a material balance oxygen uptake can be distributed with respect to maintenance and growth according to a very simple model[64]. The values of the constants c_m and $Y_{x/s}$ necessary for biomass calculation can then be determined by using the appropriate evaluation procedures from experiments in which oxygen uptake and biomass concentration are independently measured[9, 66, 68, 97].

2.3 Process Control and Optimization

Owing to very encouraging results and adequate economic justification, computer control is widely used in batch chemical processes[38]. Hence, several control strategies and systems have been also developed for large-scale processes[27, 28, 93]. Triggered by the introduction of highly instrumented bioreactors[22, 33, 34] for biochemical and microbial research, control structures and algorithms as well as optimization procedures have been tested several times[2, 40, 42, 59, 61, 73, 74, 82, 96].

In process control applications control tasks performed by the computer may first include "On/Off-control". Thus, the timing of several events before, during and follow-

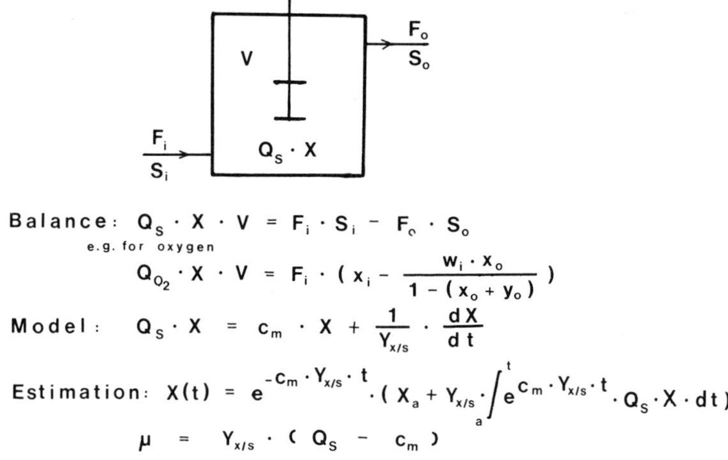

Fig. 1. Indirect estimation of biomass and specific growth rates. After computing the amount of assimilated substrate by balancing methods, it is distributed to maintenance and growth according to a simple model proposed by Pirt[64]

ing the microbial cultivation may be performed. Examples of this type of process control include:

heating for sterilization and cooling[28, 52, 93]; sampling and sample preservation for off-line determination of different process parameters by physical, chemical, and enzymatic methods[35, 84, 85]; switching of sample streams from several reactor units to shared analyzers[43, 49]; stopping the process and isolating culture liquid, biomass, and/or products[28, 93].

For further control tasks two basic strategies are currently available: "Direct Setpoint Control" (DSC) and "Direct Digital Control" (DDC[24, 38, 50]). Conventional analog control loops make it possible to maintain environmental parameters at previously defined values (setpoints). In monitoring control the computer provides the setpoint sequencing and timing for the analog controller, whereas in direct digital control, the computer directly controls the position of the final control elements. The advantage of DSC is that in the case of computer failure, control may be returned to the local analog element with manual override; on the other hand, DDC permits greater flexibility and more precise representation, since control responses can be modelled as algebraic functions: i.e., costly and time consuming redesign of controller hardware is not required.

At first, computer-aided process control enables the construction of control loops for directly accessible environmental parameters, thus implementing sophisticated control strategies by using selected combinations of proportional, integral and differential (PID) control algorithms. Process indicators or state variables, obtained by data reduction, may likewise be used in control loops[40, 42, 59, 61, 86]. When using these variables (interactive control), the fact will have to be taken into consideration that the change of even one environmental parameter influences the values of several other variables, so that their readjustment will be necessary. This adjustment must be carried out according to the complex physiological needs of the cultivated cells, hence, a detailed knowledge of microbial physiology is of utmost importance. These control operations may act on several levels for instance connected in cascades. The target values of environmental variables are maintained by means of conventional control elements (DDC or DSC), but the setpoint of each control loop is altered according to the state of the culture by taking into consideration the interactive effects of the system elements.

The algorithms of interactive control can be developed most effectively from on-line computer-aided experiments while observing the response of a culture to a sudden change of one variable (perturbation testing[25, 61, 63]). Results of modelling experiments may likewise be used after process identification and parameter estimation[9, 11].

Computer-aided interactive control was successfully implemented in experiments using the computed $k_L \cdot a$ - value to alter agitation speed and air flow rate in order to maintain the predetermined concentration of dissolved oxygen. A physiological process parameter calculated on-line, in particular the RQ, was utilized to control the feed rates of carbohydrate and nitrogen sources in a cultivation of *Candida utilis*[61], or to control the molasses feed rate in the case of fed-batch cultures of baker's yeast, respectively[86, 89].

Optimization procedures are typically designed to increase the concentration of some extracellular product, to enhance particular metabolic functions, or to increase the production of biomass. Due to the very complex nature of microbial systems, experiments concerning on-line optimization can scarcely be found in recent literature.

The ultimate aim of several experiments performed using a continuous culture system (chemostat, turbidostat) was to obtain the appropriate values of environmental parameters (e.g., temperature, pH, dilution rate, etc.) for optimum biomass productivity[37, 82]. Search procedures applied for maximum productivity were the simplex technique, or combinations of several methods, e.g., Powells method and the technique of Hooke and Jeeves[82]. Calculation of the values of environmental parameters for the next optimization step is feasible if steady state conditions are reached; hence, statistically confident values for dry weight concentrations are needed, i.e., according to the dilution rate every 2–10 h. In these types of experiments not only does the computer perform data acquisition and reduction, but also calculates the values of further environmental parameters and changes the setpoints of the different control loops according to the calculated values.

The attainment of a maximum (e.g., productivity, growth rate, product formation rate, etc.) sometimes corresponds to extremes in physiological process indicators. In this case, the optimum values for environmental parameters may be gaind by detecting the extreme values of process indicators. For instance, pulse testing of *Candida utilis* continuous culture by altering the C/N-ratio resulted in the definition of an optimum C/N – value at which maximum growth rate corresponds to a minimum RQ[61].

3 Microelectronics and Microcomputers

The progress made in the development and construction of electronic switching and amplifying elements in the last two decades has not only profoundly increased the efficiency of electronic devices, but above all has enabled the construction of digital systems (computers) based on Boolean logic in large quantities and has economically justified their widespread use due to low production costs.

Due to the methods of photolithography and solid-state diffusion, microelectronic systems in integrated form can be produced today from a single silicon chip at low costs. Such integrated circuits are generally used as electronic building blocks having different functions: e.g., counter, decoder, comparator, register, flip-flop, latch, etc. The improvement of the performance of microelectronic elements has been decisive for the development of "high end" or "mainframe" computers – the largest and fastest machines –; it has been even more important for the development of "low end" computers – the smaller and slower machines. Many tasks do not require the great processing capability, flexibility and speed of even a small mainframe computer system; they may be solved using small computer systems (minicomputers, microcomputers) able to execute comparatively simple computing functions at lower costs.

3.1 Computer Elements

The block diagram of a typical computer (Fig. 2) incorporates the following elements:

1) *Central Processing Unit* (CPU). This unit performs arithmetic and logic operations based on binary data in accordance with previously stored instructions. Its dimension and efficiency depends on the length of the "word" (information) to be processed (4, 8, 16 bits). The CPU consists of a control unit to interpret instructions from the stored programme, the arithmetic and logic unit (ALU) to perform arithmetic and logic operations, and several registers which serve as easily accessible memories for data, instructions and addresses for example.

2) *Memory Devices.* These serve for the binary storage of instructions (programmes) and data and determine the capacity of the computer with respect to the amount of data and instructions. The storage mode as well as the construction and arrangement of memory elements have a great influence on the speed of operation as well as on the storage element density and memory system price. High speed memory (50–200 ns access time) are constructed predominantly in bipolar semiconductor technology; however, they are limited in capacity. Memory elements comprising field effect transistors – Metal-oxide-semiconductor (MOS) – are relatively slow (~1 μs), but they cost less because they can be packed quite densely (2^{16} bits per chip). *Random Access Memory* (RAM) are storage devices frequently used by the CPU to store and retrieve information (READ/WRITE), and predominantly contain variable programme parts and data. The information in *Read Only Memory* (ROM) is usually permanently fixed; ROM comprises programme parts repeatedly needed (e.g., compiler or interpreter programme) or adapts the computer to a specific purpose, i.e., to a user programme.

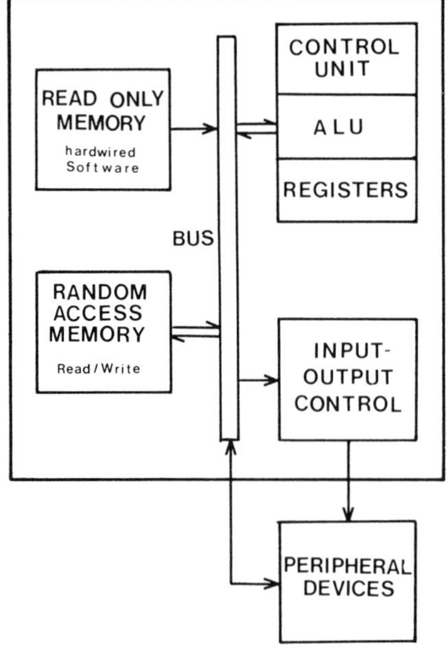

Fig. 2. Computer elements. The central processing unit (CPU), different memory devices (ROM, RAM) and elements of input/output control are connected together *via* a bus system

3) *Input/Output - Elements.* These are devices necessary for communication between the computer and the connected on-line system, or between the computer and the operator. On one hand, these elements determine the extent to which the process is acquired and controlled; on the other hand they affect ease and speed of information exchange. Existing facilities for data input use keyboard (teletype, terminal), input systems for binary stored information (paper tape, Hopper card, magnetic tape or disk) and measuring devices with digital data output. Data output may be presented optically [light-emitting-diodes (LED) display, cathode ray tube (CRT)], graphically (plotter, hard copy) or in printed form (printer, type-writer, teletype) to inform the operating personnel. Data output can also be additionally carried out for the purpose of data conservation and process control (digital control devices). The different peripheral devices are linked *via* a bus-system which consists of several signal lines, control lines and management lines.

Table 1. Classification of computers[79)]

System	Number of logic chips	Memory capacity k byte	CPU cycle time ns
Large mainframe computer	100,000	4,000–8,000	10– 100
Small mainframe computer	20,000	128	200– 500
Minicomputer	5,000–10,000	16– 64	300– 900
Microcomputer	1,000– 2,000	1– 10	2,000–10,000

Table 2. Mass storage devices of microcomputers

System	Capacity k byte	Mean access time s
Tape cartridge (cassette)	80–300	6 –19
Diskette (floppy disk)	100–300	0.3– 1
Mass memory (disk)	2,400	3

3.2 Minicomputers

The construction of moderately priced minicomputers by down-scaling mainframe computer systems has enabled on-line processing in the laboratory. In the minicomputer mode the computer remains a distinct entity being designed as a general-purpose unit but capable of being interfaced for a particular purpose. Its operation is self-contained with mass storage and terminals being directly available. The actual operation may be off- or on-line, but it is possible to interrupt operation if necessitated by the experiment. Coupling of minicomputers offering the inherent advantages of mainframe systems to one or several bioreactors has been repeatedly described (Table 3).

Table 3. Survey of some computer-coupled microbial cultivation systems in laboratory scale

Institution	Computer type capacity (word size) language	Vessel Working volume [l]	Number of sensors	Scanning cycle [min]	Ref.
Fermentation Design Inc. Bethlehem, USA	PDP–11/20 32 k (16 bit) FORTRAN	270	42	5	57–61)
Karolinska Institutet Stockholm, Sweden	HP 2100 A 12 k (16 bit) FORTRAN	3	8	0.5	82)
INRA; Station de Génie Microbiol., Dijon, France	T 2000 12 k (20 bit)	10	6	1 –60	15)
GBF Braunschweig, Fed. Rep. Germany	PDP–11	200– 500	16	15	43–45)
CHEMAP AG. Männedorf, Switzerland	TEK 31 32 k Byte	14–1,500	10–120	–	55)
Univ. Pennsylvania; Dep. Chem. & Biochem. Engineering; Philadelphia, USA	PDP–11E10 32 k (16 bit) ASSEMBLER	70	16	1–10	4)
MIT; Dep. of Nutr. & Food Science Cambridge, USA	PDP–11/10 24 k (16 bit) BASIC	14	16	0.02[a]	86, 87)
Univ. Techn. Scienc.; Inst. Agricult. Chem. Technol. Budapest, Hungary	EMG–666 8 k Byte	5– 10	20	5	37)
Univ. Technol.; Inst. Biochem. Technol. & Microbiol. Vienna, Austria	HP 9830 A 12 k (16 bit) BASIC	5– 50	12	1	30–32)
Univ. New South Wales; School Biol. Technol. Sidney, Australia	Microprocessor National SC/MP 256 RAM, 512 ROM (8 bit)	5	2	0.5–0.002[b]	89)

[a] Computation of the mean every 15 min [b] Computation of the mean after 511 readings

3.3 Microcomputers

The progress made in microelectronic technology has furthermore made it possible to produce a central processing unit of a computer together with its associated circuitry in form of a single silicon chip (microprocessor). By adding various kinds of chips to provide timing, programme memory, random access memory, interfaces for input and output signals and other ancillary functions, e.g., priority, one can assemble a complete computer system on a board (microcomputer[80)]). By using appropriately dimensioned chips, custom-tailored computer systems for particular needs and of adequate capacity can be constructed. Generally the number of assembled chips determines the size, features, efficiency and price of a microcomputing system; a simple module or breadboard system costs approximately US $ 100, whereas a Full Development System (FDS) with suitable peripheral devices (CRT, printer, mass memory, ROM-programmer) costs approximately US $ 15,000.

Although originally designed for low-end, large quantity application such as traffic light controllers and cash registers, the microcomputer was soon used for more sophisticated purposes. The actual inherent scope of this unit is still being discovered and new modes of application are continually being developed. The low cost of mass-produced microprocessors and other system components resulted in the use of microcomputers both at the low end where computers had been previously considered to be "overkill" and at the middle level where minicomputers have been used.

For real-time processing the microcomputer is now frequently incorporated in highly specific machines and instrumentation units, e.g., in CRT-terminals, spectrophotometers, polarographs, chromatographic systems (gas chromatographs, high performance liquid chromatographs), etc. This application might be termed *"in-line"*, since, to the operator the computer is indistinguishable from the instrument. The operator communicates with and directs the operation of the computer but does not program the computer; this function has been already performed at the factory by storing the programme in unalterable memory (ROM).

The relative unalterability of the user programme and the integration to a compact apparatus make such a microcomputing system look like a powerful and "smart" instrument. In order to obtain sufficient flexibility, it has several times been attempted to fix the information laid down in the ROM circuitry during manufacturing by other means. This has resulted in the development of programmable (PROM, e.g., fusible-link memory) and erasable-reprogrammable (EPROM, e.g., optically erasable or electrically alterable memory) "Read Only Memory" chips. Essential prerequisites for programming such a chip are the presence of an appropriate device (PROM – Programmer) on one hand and a comprehensive knowledge on the other hand since the programme must be written in binary machine code or sometimes in assembly language. This fact was an obstacle to the widespread use of low-cost simple microcomputer systems in chemical and biological laboratories.

Another attempt to increase the flexibility of microcomputers resulted in the development of programmable desk-top calculators which rank among the simple full development microcomputing systems. Such a calculator frequently incorporates a printer, a mass memory device (tape cassette, diskette), an alphanumeric display and a keyboard. Furthermore, it is assembled as a self-contained and compact entity. By increasing the capacity of RAM and implementing ROM as programme interpreter, a versatile micro-computing system is formed having almost the features and capacity of a minicomputer. For example, the capacity of RAM chips can amount to 62 K bits today. Since only little knowledge of computer programming is necessary and since the cost of such a system is reasonable, programmable desk calculators have, therefore, frequently been used in on-line systems in chemical and physical laboratories. The slow speed of operation may sporadically involve problems, in particular, if various computations must be performed based on a multitude of data.

3.4 Distributed Computer Network (Multiprocessor Systems)

The microcomputer represents truly low-cost computing and its economics are so compelling that several dedicated microcomputer modules can be teamed together thus forming a distributed computer network. Such an intelligent multiprocessor system may include minicomputers offering the possibility of mass storage and higher level programming language or several "in-line" systems dedicated to specific tasks. It resembles in some respects the human nervous system. The nervous system is a network of sensors and microprocessors (ganglia) connected by data links to a central computer (brain). Thus it is possible to do 99% of what is done without "thinking", that is, without involving the highest control center. In a distributed computer network system a similar phenomenon may be observed; the central computer merely orders the performance of an operation but does not actually inform each satellite computer concerning each step nor how to execute it. This greatly reduces the amount of data that must be transmitted over data links and handled by the central computer, thus enabling the computer to perform other more important tasks.

4 Microcomputers Coupled to Bioreactors

The possibilities and advantages of coupling a computer to one or several highly instrument bioreactors have been repeatedly described in detail. However, only the introduction of low-cost and efficient minicomputers rendered their use for pilot-plants and larger laboratory systems economical since in these cases the financial investment for the on-line computer amounted to a fractional part of the cost of the whole cultivation system. A reverse cost situation characterizes bench-top cultivation systems frequently used in smaller research laboratories. This fact prevented computer installation at this level. If access to large mainframe computing systems is available, an adequate priority to control microbial experiments is attainable only with difficulty.

The introduction of microprocessors has provided the financial basis for on-line computer-coupling to bench-top and similar bioreactor systems. The cost reduction owing to the use of microprocessors is compensated by the fact that programming of simple, cheap microcomputers can only be carried out sporadically by biologists or technicians. This personnel predominantly received solid and to some extent detailed training in chemistry, biology and/or process engineering sciences, but has only a limited knowledge of electronics and computer sciences. With the exception of some persons, this personnel is usually not qualified to assemble microcomputers from several chip modules or to design the software in binary machine code. This is obviously the reason that in spite of practically no cost limitations the widespread use of microcomputers in biological research laboratories is not encountered.

The utilization of computer-aided, highly instrumented bench-top bioreactors will presumably be possible to a greater extent when several manufacturers of laboratory cultivation systems offer intelligent in-line systems at a reasonable price. Such a fully integrated compact system, fitted together from several functionally adequate and compatible units, will probably be available in the near future. These systems will be already programmed *via* hardware design and will have a possibility for defining variable factors *via* a numeric keyboard; such a system will perform specific operations upon a single key stroke.

In order to interface existing microbial cultivation facilities which sometimes differ largely from one another in design and construction to the computer without any expert available for providing the necessary electronic circuitry and computer programmes, the following version is practicable: During the last few years several types of low-cost, programmable desk-top calculators with a large scope have appeared on the market. These are not only designed for commercial applications, but also to handle technical or scientific problems. These calculators, considered to be compact devices of a microcomputer nature are programmable in a high level language. As a result of the advantageous cost situation and the simplicity of programming, the coupling of such a calculator to existing highly instrumented bench-top bioreactors may be regarded as being optimal from the viewpoint of the present situation in the field of microcomputing systems. The possibilities for the practical application of programmable desk calculators have vastly increased during the last few years as a result of sufficiently high operation speeds on one hand and because of the extension of memory capacity, i.e., ROM as well as RAM on the other hand. Thereby, computer systems have been developed having an efficiency approximately comparable to that of minicomputers.

4.1 Hardware Configuration

A possible configuration of a computer-aided system for data acquisition and process control used in microbial bench-top cultivations is schematically shown in Fig. 3. The various peripheral devices necessary for the numerical evaluation of environmental parameters and for the realization of control signals are connected to a computer or a programmable desk-top calculator by appropriate interface units. As regards to the different individual hardware elements of such a system, the following may be stated:

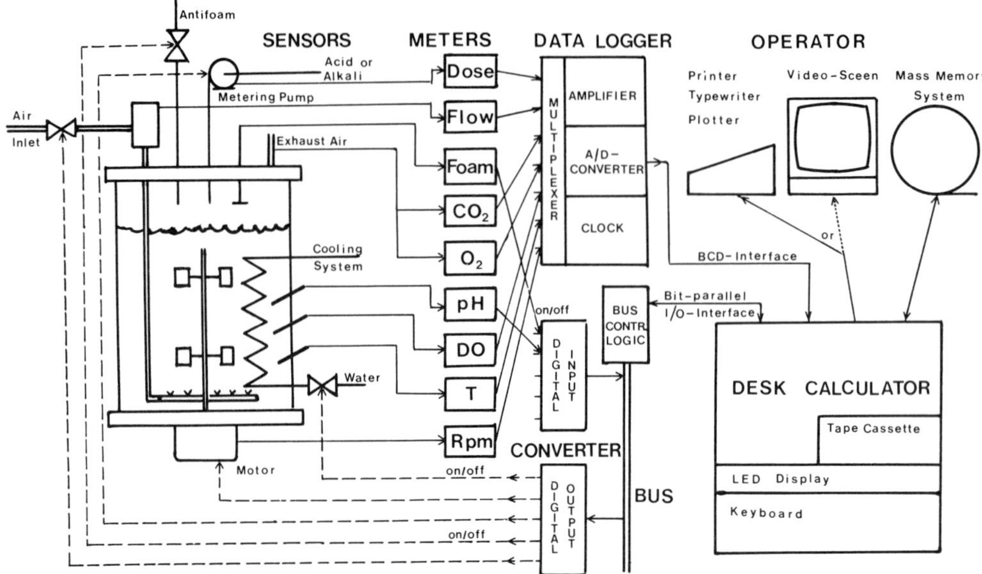

Fig. 3. Schematic diagram of hardware configuration of a system for data processing and process control in microbial cultivations. The bench-top bioreactor is coupled to a programmable desk-top calculator *via* several peripheral devices and interface units

4.1.1 Microcomputer (Desk-top Calculator)

As a central unit for controlling the different procedures and performing complex arithmetic and logic operations the following subsequent requirements should be met:

1) *Availability of large user memory.* Simple data acquisition and reduction can be performed with 4 K Bytes RAM capacity; usually an adaption of memory size to the extent of user programmes and to the number of data to be stored during operation is necessary. When required, memory capacity may be sometimes expanded afterwards by plug-in modules.

2) *High level easy-to-learn and easy-to-use programming language.* (Fig. 4) Compiler or interpreter for FORTRAN, BASIC or PL-1.

3) *Simplicity of programme corrections and changes.* Quick direct access to programme parts *via* display or video screen and the possibility to perform corrections and changes quickly without time consuming assembly run of changed programmes.

4) *Versatile editing capability and flexibility of input/output functions.* Possibilities to interface several peripheral devices in order to obtain graphic or typed reports for documentation on the progress of the experiment (printer, type-writer, CRT-display with hard copy unit, plotter), and systems for data input and output (measuring devices, relays, stepping motors) permitting different modes of data transfer (parallel, serial, asynchronous) directly or *via* an external bus-system.

```
5000 REM CALIBRATION OF pH
5010 REM DEVICE: Radiometer PHM 25    ;RANGE: 0 - 7.78
5020 E[1]=7.78-A[N,2]/1000

5100 REM CALCULATION OF AIR FLOW RATE
5110 REM DEVICE: Meter (Bach)     ;RANGE: 85 - 150 l/h
5120 REM F1: Volume of Ferm. Liquid (l)
5130 REM OUTPUT (l/l*h)
5140 F1=3.15
5150 X=A[N,13]/1000
5160 E[2]=(81.45+6.938*X+20.895*X*X)/F1

5200 REM CALIBRATION OF DO
5210 REM DEVICE: Radiometer Tox 40    ; RANGE: 0 - 150% of Sat.
5220 REM A1:Value for 100% of Sat.; A2:Value for 0% of Sat.
5230 A1=96
5240 A2=2.6
5250 E[3]=(100/(A1-A2))*A[N,3]

5300 REM CALIBRATION AND CORRECTION OF EXHAUST GAS ANALYSIS
5310 REM DEVICE:Siemens Ultramat 1   ; RANGE: 0 - 5 Vol% CO2
5320 REM DEVICE:Siemens Oxymat 2     ; RANGE: 20.9 - 11 Vol% O2
5330 REM Corrections according to FIECHTER/MEYENBURG
5340 REM OUTPUT:   OUR (mMol/l*h);   CER (mMol/l*h)
5350 X=A[N,4]/10000
5360 C1=(-0.0085+3.865*X+1.112*X*X)-0.03
5370 O1=A[N,5]/10000*10.6
5380 C2=C1+((79.07/(79.07+O1-C1))-1)*(C1+0.03)
5390 O2=O1-((79.07/(79.07+O1-C1))-1)*(20.9-O1)
5400 E[4]=1000*C2*E[2]/(100*22.416)
5410 E[5]=1000*O2*E[2]/(100*22.416)

5500 REM CALCULATION OF RQ (for O2,CO2 > 0.05 Vol%)
5510 IF C2>0.05 AND O2>0.05 THEN 5540
5520 E[6]=0
5530 GOTO 5700
5540 E[6]=E[4]/E[5]

5600 REM CALCULATION OF Kla  (JEFFERIS)
5610 REM O3: Solubility of O2 in water (mMol/l)
5620 O3=0.242
5630 E[7]=E[5]/(((20.9-O2)/20.9-E[3]/100)*O3)
```

Fig. 4. Parts of a programme written in BASIC and used at a desk-top calculator (Hewlett Packard HP 9830) for computer-controlled microbial cultivations. Algorithms for the calibration, correction, and computation of environmental parameters and state variables as pH, air flow rate, dissolved oxygen concentration, effluent air composition, RQ and $k_L a$ are presented

5) *Adequate capacity and rapid access to mass storage systems.* Internal and interfacing possibilities for external mass storage systems for data and programmes, characterized by adequate capacity and mean access time (Table 2).

6) *Priority interrupts.* In the case of unforeseen events these result in the computer to promptly starting definite programme parts. This is "hardwired" in recently developed desk-top calculators, whereas with "old timers" a similar effect can possibly be obtained with time lag by appropriate software programming, e.g., programme branching, triggered by a definite input signal.

4.1.2 Interface Units for Peripheral Devices

In order to accomplish data transfer from peripheral devices to the computer or vice verse, interface units provide an adaption or conversion of the external digital information to a form which can be processed by the computer. The following units are frequently used:

1) *BCD - Interface* - For interfacing devices with Binary Coded Decimal (BCD) output; frequently 8-10 digits may be transmitted with one interface card.

2) *Bit Parallel Input/Output Interface* - Depending on the computer system, binary signals of 8, 12, or 16 bits are accepted or sent by the computer.

3) *RS-232-C Interface* - Provides bit serial communication between the computer and asynchronous EIA RS-232-C devices, such as data terminals and modems; data transfer to a mainframe computer system is possible.

Direct connection of peripheral devices to a desk-top calculator is only possible for a few (3-4) systems. In order to interface a large number of external units, either the incorporation of a I/O - Expander (up to 15 peripheral devices) or the installation of an external bus-system (IEC - Bus) is necessary. When using a bus-system, the number of devices connected *via* interface units depends on the bus structure and size. The bus structure and/or size frequently correspond to the length of the binary coded address; e.g., in the case of a 8 bit word, a total number of 2^8 (= 256) devices may be interfaced.

4.1.3 Process Periphery

4.1.3.1 Data Input Periphery

4.1.3.1.1 Analog Signals

Analog signals (current, voltage) originating from sensors must be converted into digital (binary) form by Analog/Digital - converters. To obtain a resolution of 0.1% of the respective measuring range which is sufficient for studies of microbial cultivations, a $3\,^1/_2$ digit multimeter will meet the demands. When using signals in mV - ranges for transmitting, it is necessary to amplify these signals in order to avoid induced electric noise.

In the last few years there has been a tendency towards replacing analog by digital readouts. This tendency has been initiated by the development of integrated microelectronic circuit elements. In other words, the pointer of the traditional voltmeter has been replaced by a digital display. The number of measuring devices for process parameters equipped with digital readout is permanently increasing. Frequently, the digitally displayed value may easily be converted into a binary coded output signal (BCD - code, Octal - code) by incorporating special electronic chips which may sometimes be already integrated. The connection of such a measuring device directly to the computer *via* an interface card or an external bus system resembles the use of one A/D - converter for each analog signal and has inherent advantages. The maximum scanning rate achievable with such a system is mainly related to the operating speed of the computer and interface system. Depending on the system applied, a maximum number of 20–150 values might be acquired in one second by a desk calculator controlled system. Furthermore, each device might be addressed separately so that the sequence of parameters is determined by the software used.

In the case of analog signals originating from several sensing systems and in order to reduce the need of rather costly A/D - converters, a commonly applied technique is to multiplex the electric signals by means of multiplexers (scanners). Since extremely quick channel switching needs not necessarily be performed, relay multiplexers will fully meet the demands for data acquisition in microbial cultivation experiments; for instance mercury wetted "REED" - relays have proven to be adequate. In this case, the scanning rate depends not only on the operating speed of the computer system, but is in addition a function of multiplexer switching time as well as of the time required to range the digital multimeter and to settle the readings. By using such a system, data may be acquired at a maximum rate of 1–20 readings per s.

4.1.3.1.2 Digital Signals

Direct digital signals originating from the position of switches or from contact closure (relays) are stationary over a period of time, i.e., they are static signals; one switch or relay corresponds to one bit of information. In general the computer takes over a group of bits frequently corresponding to the length of the binary word and identifies the changed position by comparison with the sequence of bits previously stored. The expansion of digital peripheral devices for "ON/OFF" - inputs occurs, therefore, in units identical to the word length; i.e., series of 8, 12 or 16 contact inputs form one unit.

For the estimation of different process parameters (e.g., rotational speed, mass flow) a pulse counting method is preferably used. This may be performed either with special pulse counter input cards the content of which is periodically read and zeroed by the computer, or directly in such a way that the computer sums up active positions of a flip-flop over a given period of time. In order to achieve sufficient accuracy, a very high scanning rate is of utmost importance for this procedure.

4.1.3.1.3 Identification of Measurements

In order to index performed measurements temporally and locally, the acquired value of a process parameter must be furnished with additional data describing time and site of the measurement. For the real time identification a digital clock is needed either in the form of a self-contained compact device or consisting of several microelectronic modules for example by combining a frequency reference unit which generates pulses of fixed frequency and a pulse counter module which sums up the generated pulses.

Binary information identifying the external device whose signal is just transferred to the computer may be gained either *via* software, e.g., from the value of the address variable at the moment when the INPUT-statement is executed, or from status signals and switch positions (diode matrix) in case of directly inaddressible subsystems such as multiplexers.

For purposes of safety and noise immunity the peripheral circuitry should be isolated from the I/O – lines of the computer, e.g., *via* flying-capacitor energy transfer circuits or photo isolators.

4.1.3.1.4 Peripheral In-line Systems

For several applications intelligent microprocessor-controlled, in-line systems have been used with great success. By interfacing such a device to mini- or microcomputers, a distributed computer system is formed. This is of great importance if the central computer has no adequate memory capacity or has a slow operating speed.

Data Logger: "Data-Loggers" are self-contained, compact devices used to acquire and record process parameters. They include several modular devices to provide multiplexing, amplification, A/D – conversion and dating of sensor signals. They also possess facilities for data output and storage (printer, paper tape, cassette) and contain an in-line microcomputer (memory capacity 1–4 K Bytes) for system control. The intervals between consecutive scanning cycles, the parameters of which ought to be recorded (channel number), and limit values (upper and lower limits) must be defined *via* a simple numeric keyboard. A simple (linear) calibration, e.g., for thermocouples is possible. In high-performance data loggers, sophisticated linearization algorithms can be implemented by programmable ROMs. Whereas in simple low-cost systems only the whole scanning cycle may be initiated, direct addressing is possible with expensive systems. Connection of the data logger to a central microcomputer distributes the operations performed. The data logger provides the information on directly accessible environmental parameters as well as the timing of the data acquisition cycle whereas the microcomputer performs complex computations of directly inaccessible state variables and controls the cultivation process on the basis of sophisticated control algorithms.

Analytical Devices: For the majority of substances dissolved in the nutrient medium of a microbial culture no adequate sensors exist, nevertheless, in the case of utmost urgency some parameters may be determined by complex chemical and physico-chemical techniques, e.g., autoanalyzer techniques, gas chromatography, high-performance liquid chromatography, etc., as repeatedly described[3, 17, 88]. The coupling of such an analytical device to a process computer enables the automatization of the whole procedure including sampling and computing the results.

In the case of analytical systems only operating periodically, for examples chromatographic separation methods, the time required to separate the important compounds and to regenerate the analytical device must be as short as possible. It should not, if possible, exceed the time interval between two consecutive data acquisition cycles. Hardware size and configuration for successful computer coupling is mainly determined by the analytical problem to be solved and the methodology applied. For instance, if a gaschromatographic separation technique is used, the hardware may be influenced by the number and type of compounds to be separated and, as a consequence, by the necessary control requirements for temperature, the detecting system (thermal conductivity, flame ionization), the gas flow switching (backflushing, bypass, pre-column), and so forth. Furthermore, the computer should have sufficient capacity for measurement and control software as well as for data reduction.

The following hardware components permit connection of a simple process chromatograph[17)]: an autoranging amplifier interfaced to an A/D - converter ($3\,^1/_2$ digit) provides a digital representation of the detector output (frequency of measurements 4-15 points per s). A basic system complement includes, furthermore, several (approx. six) ON/OFF - control lines for actuating time-coded chromatograph functions and an additional capacity in RAM (2 K Bytes) for measurement and control software.

If a programmable desk-top calculator performs the task of a process computer, it must be taken into consideration that during a definite period the calculator will be fully occupied with the control of the analytical device and the interpretation of the chromatogramme so that no other operation will be controlled. Because of this limitation in operating speed, it is more effective to furnish the analytical device with its own microcomputer which controls the analytical procedure and interprets detector outputs. Only those results actually required (e.g., retention time, peak area) are transmitted to the desk calculator. Such microcomputer-controlled integrating chromatographs are manufactured as compact in-line systems at reasonable costs.

4.1.3.2 Data Output Periphery

The functions of different output devices consist of immediate information to the operator, the conservation of data from the process, and process control. Some of these functions may be performed by devices generally integrated in compact microcomputers, e.g., data storage on magnetic systems (cassette, flexible disk) and data output by printer or *via* display or video screen. Sometimes the efficiency of such an integrated system fulfills requirements only insufficiently; in this case, it is useful to connect appropriate peripheral systems frequently offered by the manufacturer of the computer. This can be easily achieved with suitable interface modules which can be simultaneously supplied.

A finished list of values of measured or calculated process parameters (Fig. 5) is obtained by means of suitable printing systems (type-writer, thermoprinter, character-printer, line-printer). A graphical representation of the experiment requires either one of the printing systems cited before (Fig. 6) or, if a better graphical resolution is desired, a plotter or a video screen with a hard copy unit.

Problems and limitations arising from the internal nonvolatile memory system for data and programme storage might be overcome by interfacing adequate peripheral systems. These differ in capacity and mean access time (Table 2). For data storage performed on successive memory segments (files), the time required for the whole storage procedure is a fractional part of the mean access time specified in Table 2, e.g., 3 s instead of 19 s on tape cassette. Furthermore, a fully developed microcomputer can serve as a terminal for a large mainframe computer system so that data may be transferred batch-wise to the high performance storage device of such a system.

Process control by a computer can be realized only after conversion of the binary coded output to an adequate electrical signal which is interpretable by the individual

EXPERIMENT NR. : A 17

Date	:	1977-07-06
Organism	:	Arthrobacter sp.
Medium	:	Dosage of 15 g Glucose
Fermentation-Volume (l)	:	3.12
Air Flow Rate (l/h)	:	112.1
Temperature (CG)	:	30
Fermenter	:	NBS Labroferm 51 : FBT : 600 RPM
Inoculum	:	Precultiv.(22 hr) on 1% Glucose

t (min)	pH	DO (%)	CO2 (%)	O2 (%)	DW (g/l)	NH3-N (mg/l)	S (g/l)	KOH (mval)
0	6.63	91.8	0.08	0.07	4.588	208	4.598	0.0
5	6.51	61.7	0.52	0.99	4.674	205	4.343	0.0
10	6.40	55.6	0.90	1.28	4.836	198	4.053	1.1
15	6.38	53.0	1.26	1.46	5.119	198	3.857	6.7
20	6.48	50.3	1.38	1.55	5.341	191	3.523	12.0
25	6.47	48.6	1.54	1.62	5.549	187	3.209	13.9
30	6.44	47.1	1.48	1.68	5.734	182	2.913	17.6
35	6.48	45.3	1.67	1.74	5.888	174	2.639	19.4
40	6.46	46.9	1.70	1.77	5.999	164	2.319	21.6
45	6.46	45.5	1.73	1.84	6.128	159	2.043	25.3
50	6.46	44.5	1.77	1.90	6.264	150	1.730	29.0
55	6.47	42.3	1.79	1.93	6.364	143	1.442	32.8
60	6.46	40.2	1.83	1.97	6.465	135	1.138	35.7
65	6.46	37.9	1.84	2.01	6.566	124	0.828	39.5
70	6.46	36.8	1.89	2.08	6.678	116	0.525	42.7
75	6.46	32.6	1.92	2.12	6.753	106	0.206	46.5
80	6.48	45.4	1.94	1.97	6.804	97	0.085	49.6
85	6.54	73.4	1.24	0.82	6.784	96	0.077	49.9
90	6.57	78.2	0.81	0.61	6.793	95	0.066	49.9
95	6.58	77.8	0.63	0.56	6.793	95	0.074	49.7
100	6.59	79.1	0.57	0.56	6.745	93	0.079	49.9
105	6.59	78.8	0.55	0.56	6.766	92	0.085	49.8
110	6.58	78.8	0.54	0.54	6.722	91	0.069	49.9
115	6.59	78.8	0.53	0.56	6.739	90	0.072	49.9
120	6.59	79.3	0.52	0.54	6.722	88	0.073	50.0
125	6.58	79.9	0.49	0.52	6.707	87	0.071	49.9
130	6.58	81.1	0.47	0.49	6.695	87	0.074	49.9
135	6.57	82.4	0.45	0.48	6.704	86	0.073	49.9
140	6.57	82.1	0.41	0.49	6.693	84	0.072	50.0
145	6.56	82.4	0.39	0.47	6.689	84	0.069	49.9
150	6.55	82.5	0.38	0.47	6.722	84	0.071	50.0

control element. The latter requires electrical signals in analog or digital form for example, votage, current or pulses. For this purpose, the binary computer output is stored prior to conversion into the desired electrical signal. The number of bits in the computer output determines, furthermore, either the number of control possibilities or the resolution of the control signal; e.g., a 8 bit output signal may be used either to control 8 different devices in ON/OFF - mode or to change one control signal in steps of 1×2^{-8} (= 1/256) of the control range. A better resolution may be obtained if the values of two or more addresses are combined to generate the control signal.

4.2 Efficacy and Limitations

Past studies with computer-aided microbial cultivations have so far involved minicomputers. Table 3 briefly summarizes some of these studies in so far as they dealt with laboratory systems. This summary shows clearly that the capacity of the RAM of the minicomputers used is comparable to that of more elaborate desk-top calculators. Even an appreciable number of signals from a bioreactor system - 10–20 sensors represent a typical number - can be processed without any insurmountable difficulties.

The frequency of data acquisition from one sensor depends on several facts as explained below in greater detail. In the course of the cultivation of microorganisms in a bioreactor, various parameters show relative slow changes, thus only requiring intervals between 5–15 min in scanning cycles. Even when using large time lags for the response of each measuring system before acquisition, there is sufficient time left to directly correct acquired data and to compute process state variables.

If environmental parameters and various state variables must be used in several control loops, short intervals between successive measurements of a parameter are sometimes needed. In order to prevent major setpoint deviations the scanning cycle ought to be initiated every 0.5–1.0 min. The necessary computations may still be performed in desk calculator-controlled systems, but their limits are reached very soon, especially when implementing several interactive control loops. The reason for this is the slow computation and execution speed (100–1000 times) as compared to minicomputers. The execution of user programmes in read/write-memory of desk calculators is interpretive in the sense that the form of the programme is not directly executable machine code, but are pointers to execution codes, symbol tables or scratch-pad memory.

Very short scanning intervals (1 s) are frequently used to achieve greater accuracy of some parameters; the multitude of values coming from a single measuring device is taken at the end of a definite period of time to calculate the mean and additionally the standard deviation. This procedure is applied in particular to staggered or noisy signals requiring the application of digital filter algorithms[8, 45, 47]. In desk calculator-aided data

◀**Fig. 5.** Listed data of directly accessible environmental parameters printed by the output type-writer (Facit 3841) of a desk-top calculator. In the experiment presented glucose (15 g) was added to a nutrient depleted culture of *Arthrobacter sp*. The values of several parameters, e.g., pH, dissolved oxygen, carbon dioxide and oxygen in effluent air, biomass, ammoniacal nitrogen, glucose, and added alkali were acquired every minute but only printed in intervals of 5 min

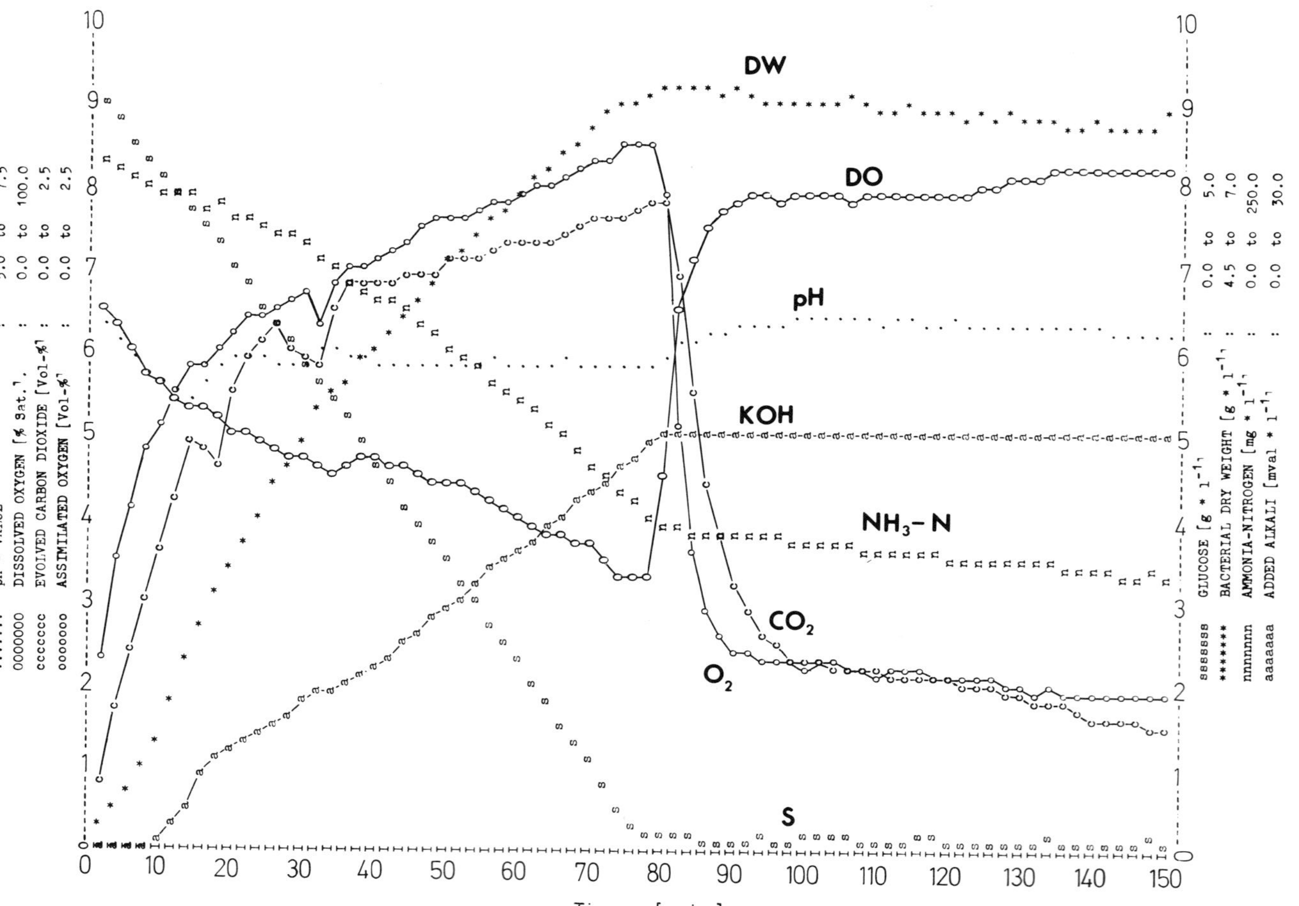

Fig. 6. Data of directly accessible parameters printed in intervals of 2 min by the output type-writer of a desk-top calculator in graphical form. Experiment and experimental conditions are already described in Fig. 5

acquisition systems, the computation of the mean and standard deviation is possible; digital filtering, however, is impracticable.

Fully developed microcomputers, e.g., programmable desk-top calculators are largely comparable to minicomputers for use in coupled cultivation systems concerning tasks performed hitherto such as data acquisition, data reduction, and process control. To what extent limitations due to the slow execution speed may occur in specific tasks cannot be stated a priori since these will be essentially influenced by the type of microcomputer used as well as by the hardware configuration and extent of the user programme.

4.3 Cost Considerations

Stating the cost of computer systems or single elements thereof, is always difficult and problematic since, on one hand, devices offered for sale never completely correspond to each other. They always differ somehow in construction, performance, efficacy, and applicability. On the other hand, the progress in electronics technology is so enormous even today that devices presented some years or months ago as the latest development are frequently no longer produced since they have already been replaced by better and cheaper systems. Despite the permanently changing price situation, it has been tried to give a survey of the current situation in 1978 in order to document the low costs of online, intelligent, and compact microcomputer (desk-top calculator) systems (Table 4). This is additionally justified by the fact that the challenge for the biotechnologist will not be to design a computer system to record data and control environmental para-

Table 4. Costs of several parts in desk-top calculator aided cultivation systems

Device	Price US $
Desk-top calculator (8 K Byte RAM, BASIC, Display or CRT, internal mass memory)	10,000–15,000.–
Memory expansion (unit of 8 K Byte)	2,000.–
Output – device (type-writer, printer, plotter, Hard copy unit)	5,000– 8,000.–
Interface unit for one analog input device (A/D–converter, interface card)	800.–
Interface unit for output elements (Interface card, converter)	800.–
Digital clock	1,000– 1,500.–
External mass memory system	
Tape cassette	3,000– 4,000.–
Flexible Disk	7,000.–
Disk	20,000.–
In-line devices	
Data logger (multiplexer, A/D–converter, clock, interface) for 10 analog signals	4,000– 8,000.–
Automatic analytical devices	
Gas chromatograph (GC)	8,000–17,000.–
High performance liquid chromatography	30,000–45,000.–
Integrator	8,000.–

meters in microbial cultivations, but to choose from among a fairly large number of readily available hardware and software systems and to apply the chosen system to the particular problems of his own plant[16].

The investment costs of a complete on-line microcomputer-aided system for data acquisition, data reduction, and process control vary greatly. This is so because computer memory capacity, number and complexity of sensors and control elements, and comfort in applicability greatly affect the price. For a simple system comprising at least a compact microcomputer or a high performance desk-top calculator, respectively, and an output device (type-writer, printer), costs of approximately US $15,000.– must be expected.

Direct connection of measuring devices to a computer or to an external bus system is economically justified for the acquisition of only a few (3–4) environmental parameters, whereas, for a multitude of measuring sites the use of multiplexer or smart data logger is regarded as most favourable in so far as no configuration with inherent limitations arises, e.g., in case of high frequency data acquisition.

A great part of the estimated investment costs must be allocated to measuring devices, transducers, and sensors. This is true in particular if complex analytical systems are incorporated. This effectively restricts the utilization of these high-performance systems to individual cases where special problemes must be solved. Due to these economic considerations, a flexible integration of hardware as well as software and analytical devices into one system is not recommended. Such units must be easily and quickly exchangeable or transferable to other bioreactor systems.

Cost reduction is possible for researchers having either more time than money or a desire to make microelectronic circuitry their hobby; integrated circuits are very inexpensive and often special purpose machines may be designed having a lower parts count than is necessitated in the case of those designed for general purpose use. In this way, existing devices might be combined and interfaced to a compact microcomputer and output elements may be constructed for process control, thus resulting in systems of adequate efficacy at very low cost[10, 25, 72].

5 Broadening Applicability

The development of low-cost programmable desk-top calculators has made it possible to use highly instrumented bench-top bioreactors as computer-coupled systems. These systems permit permanent supervision of small scale microbial experiments with a relatively small staff. The simple identification of various process states such as transition state or a steady state has resulted in the application of computer-coupled bioreactors for on-line optimization of biomass productivity in continuous culture systems. Hitherto, the optimum values for some parameters which are measured and controlled by simple procedures have been obtained in this way[37, 82].

In order to develop control algorithms for optimal process control, a detailed and well-founded knowledge of the influence of different experimental parameters on the physiological and metabolic behaviour of the microbial population is of utmost necessity. In this respect it is particularly important to be able to detect correlations between on-line process indicators easily computable, e.g., Q_{O_2} X, Q_{CO_2} X, RQ, etc., as well as events or states respectively during culture development.

For example, the depletion of an energy source such as glucose in the culture liquid at 78 min elapsed cultivation time with the organism *Arthrobacter sp.* first causes a drastic change in Q_{O_2} and DO (Figs. 6 and 7). The decline in Q_{CO_2} occurs with some time lag, thus resulting in an RQ maximum 6 min later which assumes a stationary value again after approximately 20 min. The slow down in cell growth as well as assimilation of ammoniacal nitrogen and acid formation occurs likewise with a time lag. In order to detect these events which all take place within a period of approximately 20 min, a high data frequency as well as adequate sensors must be utilized. The advantage of computer-aided data acquisition as compared to the use of a simple data logger lies in the greater number of usable measuring devices, e.g., sensors supplying nonlinear signals or displaying measurements with time delay might be used.

The success of an experiment partly depends on whether or not the experimenter receives important information during the experiment on its progress and on the basis of operations or changes performed at the right moment, he can control the cultivation so as to obtain valuable results in a quick and easy manner. The on-line graphical representation of the course of individual environmental parameters and directly inaccessible state variables is of considerable help in making those decisions. In this respect adequate output devices such as a video screen, a plotter, or a printer with graphical capabilities can be of great advantage.

For example, the experiment in Fig. 8 was conducted in such a manner: The on-line graphical printed data on ethanol concentration (EtOH), turbidity, Q_{O_2} X and RQ were used for decisions on changing the manually alterable continuous feed rate of glucose (F_s) to a fed-batch culture of baker's yeast. Figure 8 represents a typewriter printout performed at the end of the experiment after off-line data analysis, thus additionally displaying the computed values of glucose concentration (S), biomass (DW), specific growth rate (μ), specific glucose uptake rate (Q_s), and specific ethanol formation rate (Q_{EtOH}). This investigation primarily served to study in detail the phase of simultaneous utilization of glucose and ethanol which occurs below a critical specific glucose uptake rate[91, 92].

Furthermore, the almost perfect correlation between the RQ and the Q_{EtOH} was used to establish algorithms for an optimal control of glucose feed. Figure 9 shows the graphical record of an experiment testing a proportional-differential algorithm which controls glucose feed rate *via* the computed RQ. Direct digital control (DDC) was used to control the feed rate of fresh nutrient medium in the case of a fed-batch culture of *Saccharomyces cerevisiae.* A precision metering pump was controlled by modulated pulses generated by a special output device which was interfaced to a desk-top calculator (HP 9830[92]). The setpoint was altered continuously; starting with a value of 1.2, it was lowered to 1.05 to the end of experiment. This was done to prevent misleading at the begin of the experiment and to avoid in the further course of the microbial process, the ethanol content does exceed an upper limit.

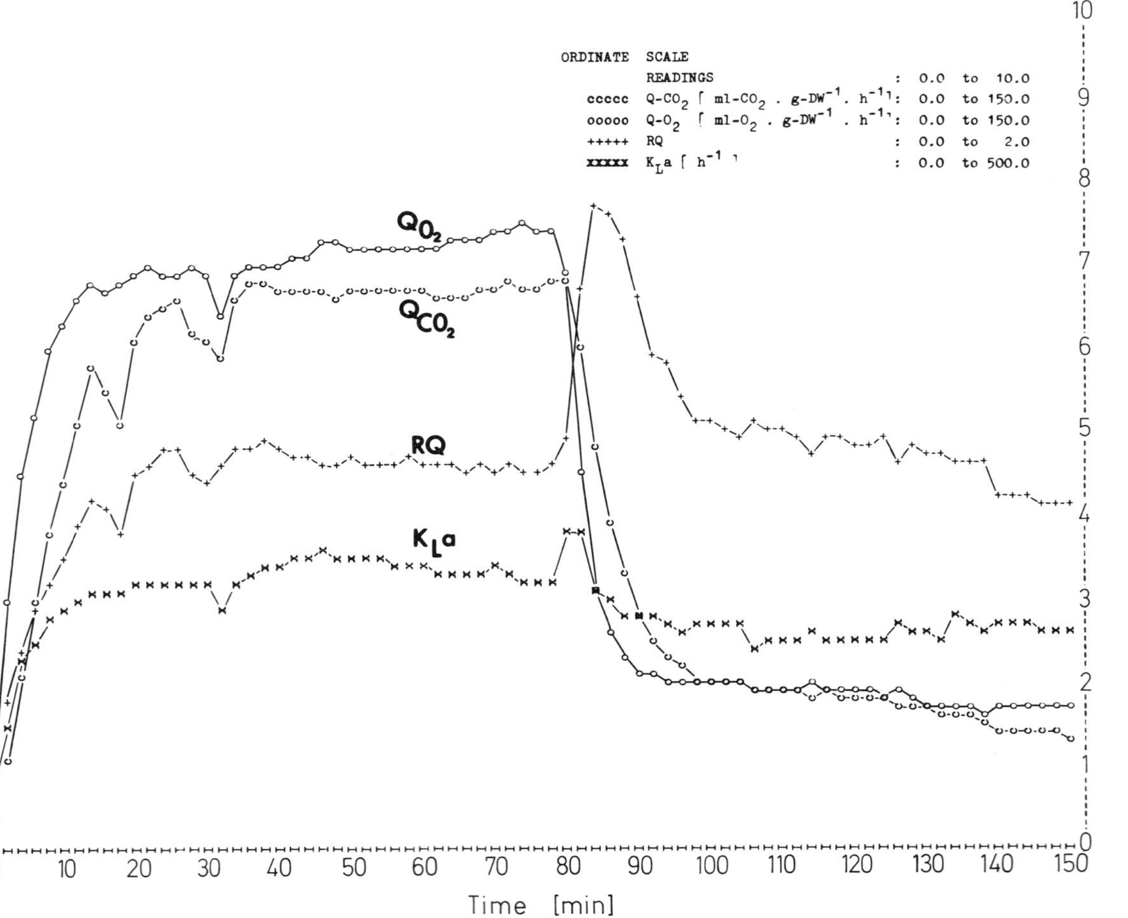

Fig. 7. Data of directly inaccessible state variables printed in intervals of 2 min by the output type-writer of a desk-top calculator. The values of Q_{O_2}, Q_{CO_2}, RQ, and k_La were computed from corresponding data on environmental parameters using appropriate arithmetic algorithms. Experiment and experimental conditions were the same as in Fig. 5

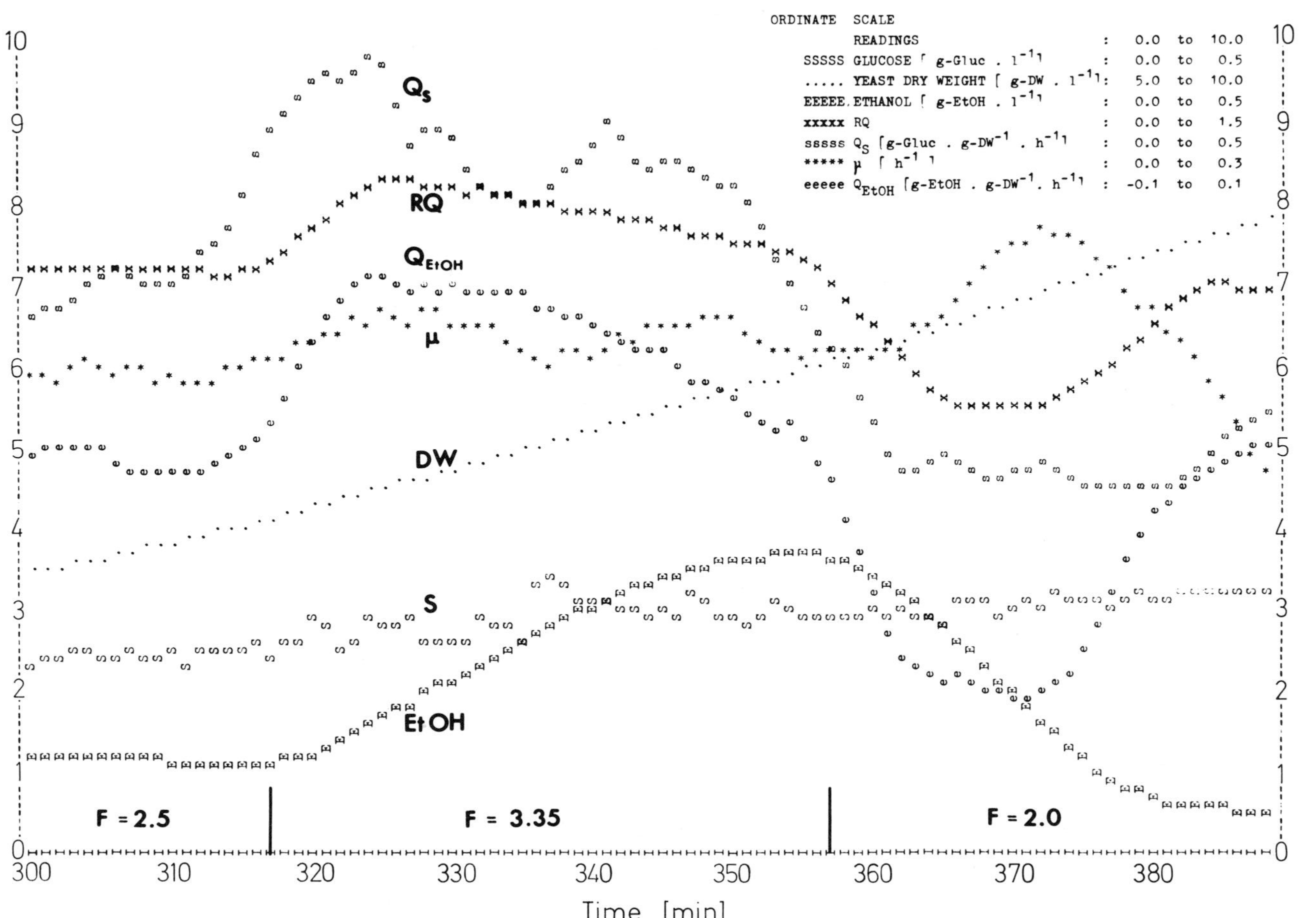

Fig. 8. Graphical type-writer print-out of directly computed data (biomass, glucose, EtOH, RQ) and off-line calculated data (μ, Q_S, Q_{EtOH}) of a fed-batch culture of *Saccharomyces cerevisiae*. Glucose feed rate [F_s ($gh^{-1}l^{-1}$)] was changed from 2.5 to 3.35 at 317 min elapsed time and to 2.0 at 317 min

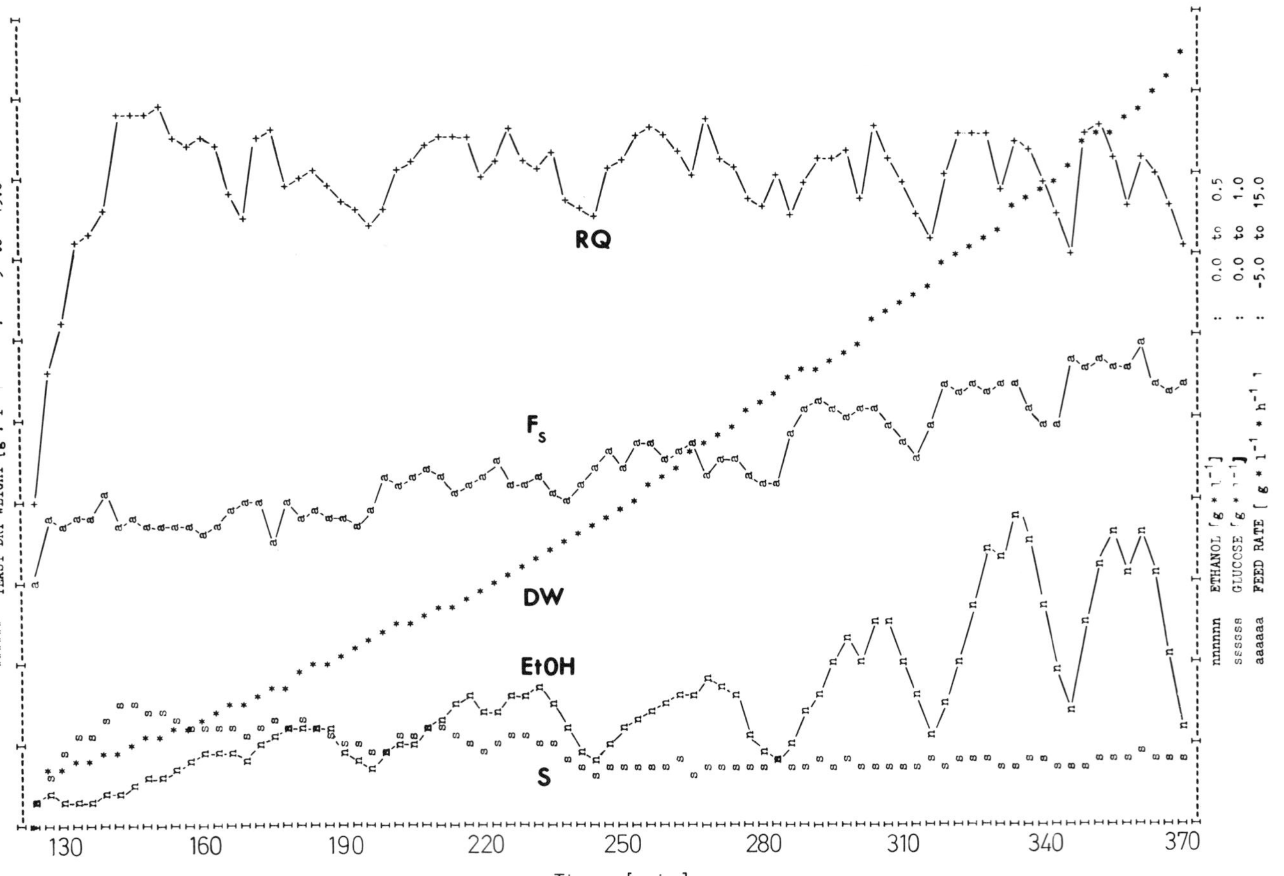

Fig. 9. Graphical print-out of some data (biomass, glucose, EtOH, RQ, and glucose feed rate) from an experiment for testing a proportional-differential control algorithm in a fed-batch culture of *Saccharomyces cerevisiae.* The RQ – value was used to control glucose feed by a precision metering pump interfaced to a programmable desk-top calculator. In order to maintain optimal growth conditions the setpoint was altered continuously

6 Concluding Remarks and Future Tendencies

The coupling of a computer to a highly instrumented bioreactor represents an effective and successful means to reveal correlations between physiological effects and physical and/or chemical environmental parameters in on-line experiments. If studies of this kind are performed with bench-top bioreactors, the utilization of desk-top calculators is economically justified. Such devices, based on microcomputer technology have capabilities and features comparable to those of minicomputers. Owing to their versatile applicability, they have gained access to the fields of commerce, engineering, and scientific research. They are about to become just as familiar to every scientifically working person as was the slide-rule until some years ago. Furthermore, the simultaneous development of high performance desk-top calculators and the simplification of interfacing peripheral devices have led to the development of versatile, low-cost, and quickly installable systems for data acquisition and process control.

Even the manufacturers of measuring devices support this tendency by developing an increased number of apparatuses with digital readout, already equipped for data transfer to a computer. If a measurement can be made only by employing complicated analytical devices, there is a trend towards the development of microprocessor-controlled versatile in-line systems, steadily increasing in capability and flexibility. Such units which are frequently adapted to communicate with mini- or microcomputers, may form satellites in a distributed computer network. As an example, either various computer-controlled analytical devices (gas chromatographs, high performance liquid chromatographs) or devices for data acquisition (data logger) may be mentioned.

High performance microprocessor-controlled data loggers, equipped with programmable read only memories (PROM) may already at the present time be adapted to serve specific tasks. In the very near future such devices might be coupled to bioreactors in order to measure a diversity of environmental parameters, to compute easily calculable state variables, and to perform simple process control (ON/OFF) tasks if desired. A reprogramming of read only memory is necessary, if the sensing systems connected thereto are changed; therefore, high performance data loggers will presumably be used only for specific applications regularly repeated.

Coupling of a simple or even an extensive data acquisition system to a computer or programmable desk-top calculator respectively, will be necessary if either the great flexibility of software such as high level programming, programme package for specific applications, or the capabilities of output devices such as video screen, plotters, or a mass memory system are required. The plummeting cost of microelectronic circuitry and fully developed microcomputers will lead to the quick disappearance of the exclusivity of computer-coupled microbial processes. The progress achievable in science and technology, will be manifested by a comprehensive and detailed knowledge of the metabolic and physiological behaviour of microorganisms and by the detection of effective control algorithms for optimal process control.

Acknowledgements

The author whishes to express his most grateful thanks to Prof. Dr. M. Röhr for helpful discussions and continuous support and also to Prof. Dr. R.M. Lafferty and Prof. Dr. J. Meyrath for their help in the preparation of the manuscript.

7 Symbols

ALU	arithmetic and logic unit
c_L	concentration of dissolved oxygen (mMol l^{-1})
c_L^*	concentration of dissolved oxygen in equilibrium with air (mMol l^{-1})
c_m	specific substrate uptake rate for cell maintenance (g–S g–X^{-1} h^{-1})
CPU	central processing unit
CRT	cathode ray tube
DO	dissolved oxygen
DDC	direct digital control
DSC	direct setpoint control
F	gas flow rate (mMol h^{-1})
F_s	substrate feed rate (g–S h^{-1} l^{-1})
h	absolute humidity (dimensionless)
K_La	volumetric oxygen transfer coefficient (h^{-1})
P	absolute pressure (Pa)
Q	specific metabolic rate (mMol g–X^{-1} h^{-1})
RAM	random access memory
ROM	read only memory
RQ	respiratory quotient (dimensionless)
S	substrate concentration (mMol l^{-1})
t	time (h)
T	temperature (°C)
V	liquid volume in reactor (l)
w	mole fraction of inert gas in air (dimensionless)
x	mole fraction of oxygen in air (dimensionless)
X	biomass concentration (g l^{-1})
y	mole fraction of carbon dioxide in air (dimensionless)
Y	yield coefficient (dimensionless)

Greek Letters

μ	specific growth rate (h^{-1})

Subcripts

a	initial conditions
i	inlet
o	outlet
s	substrate

8 References

1. Aiba, S., Humphrey, A.E., Millis, N.F.: Biochemical engineering. New York: Academic Press 1973
2. Aiba, S., Nagai, S., Nishizawa, Y.: Biotechnol. Bioeng. *18*, 1001 (1976)
3. Ames, H.S., Crawford, R.W.: Int. Labor. *1977* (3/4), 45
4. Armiger, W.B., Zabriskie, D.W., Humphrey, A.E.: In: Proc. 5. Int. Ferm. Symp. Delweg, H. (ed.), p. 33. Berlin: Inst. f. Gärungsgew. u. Biotechnol. 1976
5. Bach, H.P.: Thesis, Techn. Univ. Wien, Inst. Biochem. Technol. and Microbiol., 1976
6. Bach, H.P., Wöhrer, W., Röhr, M.: Biotechnol. Bioeng. *20*, 799 (1978)
7. Blachère, H., Jamart, G.: Biotechnol. Bioeng. *11*, 1005 (1969)
8. Bogner, R.E., Constantinides, A.G.: Introduction to digital filtering. London: Wiley 1975
9. Bourdaud, D., Foulard, C.: Proc. 1st European Conf. Computer Process Control in Fermentation. INRA: Dijon, France 1973
10. Bungay, H.R.: Biotechnol. Bioeng. *18*, 741 (1976)
11. Calam, C.T., Russell, D.W.: J. appl. Chem. Biotech. *23*, 225 (1973)
12. Constantinides, A., Spencer, J.L., Gaden, E.L., Jr.: Biotechnol. Bioeng. *12*, 803 (1970)
13. Constandinides, A., Spencer, J.L., Gaden, E.L., Jr.: Biotechnol. Bioeng. *12*, 1081 (1970)
14. Cooney, C.L., Wang, H.Y., Wang, D.I.C.: Biotechnol. Bioeng. *19*, 55 (1977)
15. Corrieu, G., Blachère, H., Geranton, A.: Biotechnol. Bioeng. Symp. *4*, 607 (1974)
16. Dobry, D.D., Jost, J.L.: In: Ann. reports on fermentation processes. Perlman, D. (ed.), Vol. 1, p. 95. New York: Academic Press 1977
17. Downer, W.: Intern. Labor. *1975* (3/4), 35
18. Emmerson, E.: 2nd Int. Ferm. Symp., London 1964
19. Erickson, L.E., Minkevich, I.G., Eroshin, V.K.: Biotechnol. Bioeng. *20*, 1595 (1978)
20. Erickson, L.E., Selga, S.E., Viesturs, U.E.: Biotechnol. Bioeng. *20*, 1623 (1978)
21. Eroshin, V.K.: Process Biochem. *12* (6), 29 (1977)
22. Fiechter, A.: Biotechnol. Bioeng. *7*, 101 (1965)
23. Fiechter, A., Meyenburg, K., von: Biotechnol. Bioeng. *10*, 535 (1968)
24. Flynn, D.S.: Biotechnol. Bioeng. Symp. *4*, 597 (1974)
25. Gaudy, A.F., Jr.: Biotechnol. Bioeng. *17*, 1051 (1975)
26. Goma, G., Ribot, D., Pourciel, J.B.: In: Proc. 5th Int. Ferment. Symp. Delweg, H. (ed.), p. 19. Berlin: Inst. Gärungsgew. u. Biotechnol. 1976
27. Grayson, P.: Process Biochem. *4* (3), 43 (1969)
28. Greiner, B.: Chem.-Ing.-Techn. *46*, 680 (1974)
29. Hampel, W., Hofbauer, P.: Zentralbl. Bakteriol., I. Ref. *240*, 310 (1974)
30. Hampel, W., Wöhrer, W., Bach, H.P., Röhr, M.: In: Proc. 5th Int. Ferment. Symp. Delweg, H. (ed.), p. 32. Berlin: Inst. Gärungsgew. u. Biotechnol. 1976
31. Hampel, W., Bach, H.P., Röhr, M.: In: Workshop Computer Appl. Ferment. Technol. 1976. Jefferis, R.P. (ed.), p. 47. Weinheim: Verlag Chemie 1977
32. Hampel, W., Wöhrer, W., Bach, H.P., Röhr, M.: Mitt. Versuchst. Gärungsgew. Wien *33*, 13 (1979)
33. Harmes, C.S.III.: Developments Industr. Microbiol. *13*, 146 (1972)
34. Harrison, D., Harmes, C., Humphrey, A.E.: Proc. 10th Int. Congr. Microbiol., Mexico City 1970
35. Harrison, D., Harmes, C.: Process Biochem. *7* (4), 13 (1972)
36. Ho, L.Y., Humphrey, A.E.: Biotechnol. Bioeng. *12*, 291 (1970)
37. Holló, J., Keviczky, L., Kirchknopf, L., Kurucz, I., Nyeste, L., Sevella, B., Szigeti, L., Veres, A.: In: Rec. Res. Chemistry (Hung.). Csakvari, B. (ed.), Vol. 39, p. 11. Budapest: Akademiai Kiadó 1978
38. Huang, M., Sonn, M.: Brit. Chem. Eng. and Proc. Techn. *17*, 507 (1972)
39. Humphrey, A.E.: Proc. of LABEX Symposium. Earls Court, London 1971

40. Humphrey, A.E.: Proc. 1st Europ. Conf. Computer Process Control, INRA, Dijon, France 1973
41. Humphrey, A.E.: Process Biochem. *12* (2), 19 (1977)
42. Jefferis, R.P.III.: Proc. 1st Europ. Conf. Computer Process Control, INRA, Dijon, France 1973
43. Jefferis, R.P.III.: In: Workshop Computer Appl. Ferm. Technol. 1976. Jefferis, R.P. (ed.), p. 21. Weinheim: Verlag Chemie 1977
44. Jefferis, R.P.III.: Process Biochem. *10* (3), 15 (1975)
45. Jefferis, R.P.III., Winter, H., Vogelmann, H.: In: Workshop Computer Appl. Ferment. Technol. 1976. Jefferis, R.P. (ed.), p. 141. Weinheim: Verlag Chemie 1977
46. Lane, A.G.: Proc. 1st Europ. Conf. Computer Process Control, INRA, Dijon, France 1973
47. Lynn, P.A.: Med. and Biol. Eng. and Comp. *15*, 534 (1977)
48. Maddix, C., Norton, R.L., Nicolson, N.J.: Analyst *95*, 738 (1970)
49. Marten, J.: In: Methods in microbiology. Norris, J.R., Ribbons, D.W. (eds.), Vol. 6B, p. 319. New York: Academic Press 1972
50. Meskanen, A., Lundell, R., Laiho, P.: Process Biochem. *11* (5), 31 (1976)
51. Metz, H.: Chem.-Ing.-Techn. *43*, 60 (1971)
52. Metz, H., Wenzel, F.: In: Proc. 5th Int. Ferm. Symp. Delweg, H. (ed.), p. 35. Berlin: Inst. f. Gärungsgew. u. Biotechnol. 1976
53. Mor, J.R., Zimmerli, A., Fiechter, A.: Anal. Biochem. *52*, 614 (1973)
54. Moss, F.J., Bush, F.: Biotechnol. Bioeng. *9*, 585 (1967)
55. Müller, F.: Process Biochem. *11* (9), 24 (1976)
56. Nyiri, L.: In: Adv. Biochem. Eng. Ghose, T.K., Fiechter, A., Blakebrough, N. (eds.), Vol. 2, p. 49. Berlin, Heidelberg, New York: Springer 1972
57. Nyiri, L.: Developments Industr. Microbiol. *13*, 136 (1972)
58. Nyiri, L.K., Humphrey, A.E.: In: Fermentation Technology Today. Terui, G. (ed.). Japan, Soc. Fermentation Technology 1972
59. Nyiri, L.K., Jefferis, R.P.III., Humphrey, A.E.: Biotechnol. Bioeng. Symp. *4*, 613 (1974)
60. Nyiri, L.K., Toth, G.M., Charles, M.: Biotechnol. Bioeng. *17*, 1663 (1975)
61. Nyiri, L.K., Toth, G.M., Krishnaswami, C.S., Parmenter, D.V.: In: Workshop Computer Appl. Ferm. Technol. 1976. Jefferis, R.P. (ed.), p. 37. Weinheim: Verlag Chemie 1977
62. Oliver, B.M.: Sci. American *237*, 180 (1977)
63. Paynter, M.J.B., Bungay, H.R.III.: Biotechnol. Bioeng. *12*, 347 (1970)
64. Pirt, S.: Principles of microbe and cell cultivation. Oxford: Blackwell Scientific Publications 1975
65. Ratzlaff, K.L.: Intern. Labor. *1978* (3/4), 11
66. Reuß, M., Jefferis, R.P., Lehmann, J.: In: Workshop Computer Appl. Ferm. Technol. 1976. Jefferis, R.P. (ed.), p. 107. Weinheim: Verlag Chemie 1977
67. Reuß, M., Piehl, H., Wagner, F.: In: Proc. 5th Int. Ferment. Symp. Delweg, H. (ed.), p. 25. Berlin: Inst. Gärungsgew. Biotechnol. 1976
68. Ribot, D.: In: Workshop Computer Appl. Ferm. Technol. 1976. Jefferis, R.P. (ed.), p. 125. Weinheim: Verlag Chemie 1976
69. Röhr, M., Hampel, W., Wöhrer, W.: In: Proc. workshop biotechnology in Austria. Lafferty, R.M. (ed.), p. 33. Graz: Inst. f. Biochem. Technol. u. Lebensmittelchemie 1975
70. Ryu, D.D.Y., Humphrey, A.E.: J. appl. Chem. Biotechnol. *23*, 283 (1973)
71. Schulz, W.B.T.: Chem.-Ing.-Techn. *43*, 67 (1971)
72. Settle, F.A., Peters, P.B.: Intern. Labor. *1976* (3/4), 31
73. Shichiji, S.: In: Biochem. and Ind. Aspects of Fermentations. Sakaguchi, K., Uemura, T., Kinoshita, S. (eds.), p. 267. Tokyo: Kadausha Ltd. 1971
74. Shu, P.: In: Fermentation technology today. Terui, G. (ed.), p. 183. Japan: Soc. Fermentation Technology 1972
75. Spruytenburg, R., Dang, N.D.P., Dunn, I.J., Mor, J.R., Einsele, A., Fiechter, A., Bourne, J.R.: Chem. Eng. (London) *310*, 447 (1976)
76. Sukatsch, D.A., Nesemann, G.: Chemie-Technik *6*, 261 (1977)
77. Svrcek, W.Y., Elliott, R.F., Zajic, J.E.: Biotechnol. Bioeng. *16*, 827 (1974)

78. Swartz, J.R., Cooney, C.L.: Process Biochem. *13* (2), 3 (1978)
79. Terman, L.M.: Sci. American *237* (3), 163 (1977)
80. Toong, H.M.D.: Sci. American *237* (3), 146 (1977)
81. Topiwala, H.H.: In: Meth. Microbiology. Norris, J.R., Ribbons, D.W. (eds.), p. 35. New York: Academic Press 1973
82. Unden, A.G., Hedén, G.C.: Proc. 1st Europ. Conf. Computer Process Control in Fermentation. INRA, Dijon, France 1973
83. Vincent, A.: Process Biochem. *9* (3), 19 (1974)
84. Vogelmann, H., Reuß, M., Gnieser, J., Wagner, F.: Proc. 3rd Symp. Techn. Microbiol. Delweg, H. (ed.), p. 215. Berlin: Inst. Gärungsgew. Biotechnol. 1973
85. Vogelmann, H., Eppert, K., Wagner, F.: Proc. 5th Int. Ferm. Symp. Delweg, H. (ed.), p. 28. Berlin: Inst. Gärungsgew. Biotechnol. 1976
86. Wang, H.Y., Cooney, C.L., Wang, D.I.C.: Biotechnol. Bioeng. *19*, 69 (1977)
87. Wang, H., Wang, D.I.C., Cooney, C.L.: Europ. J. Appl. Microbiol. Biotechnol. *5*, 207 (1978)
88. Welland, J.M., Muir, A.R.: Process Biochem. *7* (10), 24 (1972)
89. Whaite, P., Aborhey, S., Hong, E., Rogers, P.L.: Biotechnol. Bioeng. *20*, 1459 (1978)
90. Whaite, P., Gray, P.P.: Biotechnol. Bioeng. *19*, 575 (1977)
91. Wöhrer, W., Röhr, M.: Proc. 4th FEMS Symposium, p. B2. Vienna, Austria 1977
92. Wöhrer, W., Röhr, M.: Proc. 6th Int. Special Symp. on Yeasts. Montpellier, France, Chair of Genetics and Microbiology (eds.), p. S II 9. ENSAM-CRAM 1978
93. Yamashita, S., Hoshi, H., Inagaki, T.: In: Ferm. Adv. Perlman, D. (ed.), p. 441. New York: Academic Press 1969
94. Yamashita, S.: In: Ferm. Technol. Today. Terui, G. (ed.), p. 179. Japan, Soc. Ferm. Techn. 1972
95. Yoshida, T., Taguchi, H.: In: Workshop Computer Appl. Ferm. Technol. 1976. Jefferis, R.P. (ed.), p. 93. Weinheim: Verlag Chemie 1977
96. Young, T.B., Koplove, H.M.: In: Ferm. Technol. Today. Terui, G. (ed.), p. 163. Japan, Soc. Fermentation Technol. 1972
97. Zabriskie, D.W., Humphrey, A.E.: AIChE J. *24*, 138 (1978)

Dissolved Oxygen Electrodes

Young H. Lee
Department of Chemical Engineering, Drexel University
Philadelphia, PA 19104, U.S.A.

George T. Tsao
School of Chemical Engineering, Purdue University
West Lafayette, IN 47907, U.S.A.

1 Introduction 36
2 Historical Development 36
3. Principle of Measurement 38
3.1 Polarographic Electrode and Galvanic Electrode 38
3.2 Theory of Operation 41
3.3 Oxygen Microelectrodes 44
4 Design of Electrodes 47
4.1 Construction Methods 50
4.2 Electrode Metals 54
4.3 Electrolytes 55
4.4 Membrane 57
4.5 Instrumentation 59
4.6 General Design Considerations 60
5 Operation of Electrodes 62
5.1 Calibration 62
5.2 Response Time 65
5.3 Effect of Temperature 67
5.4 Effect of Liquid Film 70
5.5 Handling, Maintenance, and Other Practical Considerations 71
6 Sources of Error in Measurements 73
6.1 Errors due to Probe Characteristics 73
6.2 Errors due to Measurement Medium 74
7 Applications 76
7.1 Measurement of $k_L a$ and Respiration Rate 76
7.2 Other Applications 79
8 Conclusions 81
9 Acknowledgement 81
10 Nomenclature 82
11 References 83

Recent advances in theory, construction, operation, and application of dissolved oxygen (DO) electrodes are reviewed to assist those who use or intend to use them in such areas as biochemical engineering, microbiology, and environmental engineering. Basic operating principles of membrane-covered DO electrodes and oxygen microelectrodes are presented together with methods of construction, electrode component selection, and general design considerations. Methods of calibration and effects of temperature and liquid film on electrode performance are also discussed. Sources of measurement errors due to probe characteristics and the reaction in the liquid are discussed to illustrate some of the limitations of DO electrodes. The spacial resolution of oxygen microelectrodes in local concentration measurements is also discussed. Finally, the application of DO electrodes in measuring aeration capacity and oxygen solubility is reviewed.

1 Introduction

Since its introduction by Clark in 1956[31)], the membrane-covered dissolved oxygen (DO) electrode and its modified versions have been used widely both in research and in industry. Compared with wet chemical analysis[7)] and other methods[5)], the measurement of dissolved oxygen by the membrane-covered electrode offers several advantages: simplicity; less interference by other solutes in water; rapid, in situ measurements; and, above all, continuous measurement for real-time control of oxygen concentration in bioreactors or wastewater treatment units. Although the basic operating principles are the same, DO electrodes have been developed in different areas to meet requirements of the specific applications. Examples are:

steam-sterilizable DO probes for bioreactor applications[42)],
oxygen microelectrodes for tissue oxygen measurements[136)],
fast responding oxygen probes for respiratory gas analysis[39)],
DO probes capable of measuring trace oxygen in boiler feedwaters[43)], etc.

The wide applicability of the DO electrode can be illustrated by a number of related articles in such diverse areas as

biochemical engineering[142)],
civil engineering[122)],
microbiology[14)],
medicine[20)],
physiology[136)],
chemistry[30)],
chemical engineering[78)],
mechanical engineering[113)],
oceanography[70)], etc.

In biochemical engineering, the laboratory data obtained from DO probe measurements not only give fundamental information on microbial physiology and kinetics[14)] but also form the basis for bioreactor scale-ups[140)], production yield calculations[5)], and reactor control[126)]. Thus, it is important to understand the operating principle and some of the limitations of DO electrodes in order to effectively use them in particular applications. Reviews on DO electrodes are available by a number of authors from different areas[14, 25, 34, 37, 42, 47, 61, 64, 84, 131, 136, 142)]. This paper deals with a unified survey of the recent advances in theory, construction and application of DO electrodes to assist those who use or intend to use them in their research.

2 Historical Development

The reduction of dissolved oxygen at a noble metal surface negatively polarized with respect to a reference electrode was first observed in early 1897[136)]. In 1942, Davies and Brink used this technique for measuring local oxygen tension in animal tissues[38)]. They made various "open" and "recessed" electrodes with 25 μm platinum wire enclosed in glass insulation. These are so-called polarographic oxygen microelectrodes, which have been extensively used in tissue oxygen measurements[135)]. In these applications, the oxygen microelectrode (cathode) and the reference electrode (calomel or Ag/AgCl) are separately immersed in a test medium which contains some sort of electrolyte. The major concern in electrode design has been the fast and accurate measurement of local oxygen tension in tissues. Electrodes having tip size of 1–2 μm with 95% response time of much less than 1 s were developed[135, 149)]. Recently, these microelectrodes have been used for measuring oxygen concentration gradients in microbial slime layers[28, 66)] and concentration fluctuations at the liquid surface[27, 86, 139)].

The problem of calibration and measurement associated with early noble metal polarographic electrodes led to the introduction of the membrane-covered electrode by Clark in 1956[31)]. In its original design, both the platinum cathode and the reference electrode (Ag/AgCl) were contained in a single electrode body containing the electrolyte solution, and the entire tip of the electrode body was covered with a single polyethylene membrane. The sensor is thus completely separated from the medium to be measured by a nonconducting membrane which is partially permeable to oxygen. The Clark electrode, now the basis of many commercially available polarographic DO probes (see Table 1), could be calibrated in different liquids and, furthermore, enabled dissolved oxygen measurements in non-conducting media. The major use of the early Clark electrodes was in blood oxygen measurements[131)].

In 1959, Carrit and Kanwisher[30)] used a modified version of the Clark electrode for measuring dissolved oxygen in Chesapeake Bay water. To increase the stability of the probe, they used Ag/Ag_2O instead of Ag/AgCl for the reference electrode and 0.5 M KOH for the electroylte. Since the temperature control of the bay water was not possible, they incorporated a thermistor in their probe to compensate for sensitivity change due to temperature. The Carrit-Kanwisher probe was further developed by Carey and Teal[29)]. Hospodka and Caslavsky[65)] used such a probe in microbial processes.

A different type of membrane-covered electrode, namely, the galvanic electrode, was first developed by Mancy et al. in 1962[106)]. Noting the unstable behavior of a platinum surface as the cathode, they used silver as the cathode, and lead as the anode. Unlike the polarographic electrode, this galvanic probe did not require external voltage. The voltage generated by the silver-lead electrode pair was sufficient to cause a spontaneous reduction of oxygen at the cathode. The major advantage of the galvanic probe over the polarographic type (Clark electrode) was the long-term stability of sensitivity. However, the galvanic probe had a finite life time because of the gradual oxidation of the anode surface.

Mackereth[104)] later modified the design so that the probe could be used continuously over several months without losing stability. The improvements were in the use of a silver tubing as the cathode for an increased current output, and a massive lead shot as the anode for an increased probe life.

An autoclavable galvanic probe for *biochemical engineering* work was first described by Johnson et al. in 1964[71)] and was further improved by Borkowski and Johnson[21)]. Unlike other probes, the elctrolyte chamber was vented to withstand repeated autoclaving. Also, to facilitate the fabrication of probes in the laboratory, the cathode was made from a silver wire wound in a spiral form and a flattened lead wire was used as the anode. The Borkowski-Johnson probe became popular[42, 120)] because of its long life, ease of fabrication and relatively large current output. Brookman[24)] later described a more rugged design which showed a linear response up to 10^3 mm Hg of oxygen partial pressure. Autoclavable versions of the Mackereth probe were described by Flynn et al.[49)] and Harrison and Melbourne[59)]. The modifications included decreased cathode area and the use of different membranes and electrolytes[49, 59)]. The Mackereth probe has also been widely used for monitoring and controlling dissolved oxygen in cultivation media[60, 126)].

In the area of medicine and physiology, improvements in the DO probe design have included the response time, the spacial resolution and the flow sensitivity. The *flow sensitivity* means that the output current of the probe changes depending on liquid velocity. Normally, a DO probe requires a high liquid velocity for a proper operation. This flow dependency decreases with decrease in cathode diameter. Silver[134)] and Bicher and Knisely[18)] described miniaturized Clark electrodes with tip diameters of 2–5 μm, which exhibited low flow dependency and fast response time (95% response of less than 0.5 s). These microelectrodes have been used for local oxygen tension measurements in tissues[136)]. However, these electrodes have relatively poor stability[73)] and extremely low current output requiring careful attention to amplification and instrumentation.

Several approaches have been used to incorporate the advantages of microelectrodes in macroprobe design. Fatt and Helen[48)] and Siu and Cobbold[137)] used a number of microcathodes in a single probe body. Others[13, 72, 75)] used a thin band metal as the cathode. These probes showed relatively fast response time and low flow dependency and yet produced high current output since the total surface area of the cathode was large. Some of the commercial probes for use in deep waters employ thin band metal as the cathode[13, 72)].

Other developments in DO probes were in the measurement of trace amount of dissolved oxygen and the *gas phase oxygen measurement.* Evangelista et al.[43)] described a platinum-lead galvanic probe capable of measuring dissolved oxygen down to the parts per billion range for use in boiler feedwaters. DO probes for gas phase oxygen measurements were mainly developed in the area of physiology. The major application has been in respiratory gas analysis, which requires a fast probe response. Döhring et al.[39)] obtained a 95% response time of 20 ms in their modified Clark electrode by employing a specially preparaed 0.4 μm thick Teflon membrane. However, a 95% response time of 0.1 s could be easily obtained by using a commercially available 3 μm membrane[51)]. The accuracy of the gas phase oxygen measurement by the polarographic electrode was reported to be equal to or better than that obtained by the Scholander analysis or by the paramagnetic method[146)].

3 Principle of Measurement

3.1 Polarographic Electrode and Galvanic Electrode

When an electrode of noble metal such as platinum or gold is made 0.6–0.8 V negative with respect to a suitable reference electrode (calomel or Ag/AgCl) in a neutral potassium chloride solution, the dissolved oxygen is reduced at the surface of the cathode. This phenomenon can be observed from a current-voltage diagram, called a *polarogram*, of the electrode. As shown in Fig. 1, the current increases initially with an increase in the negative bias voltage, followed by a region where the current becomes essentially constant. In this plateau region of the polarogram, the reaction of oxygen at the cathode is so fast that the rate of reaction is limited by the diffusion of oxygen to the cathode

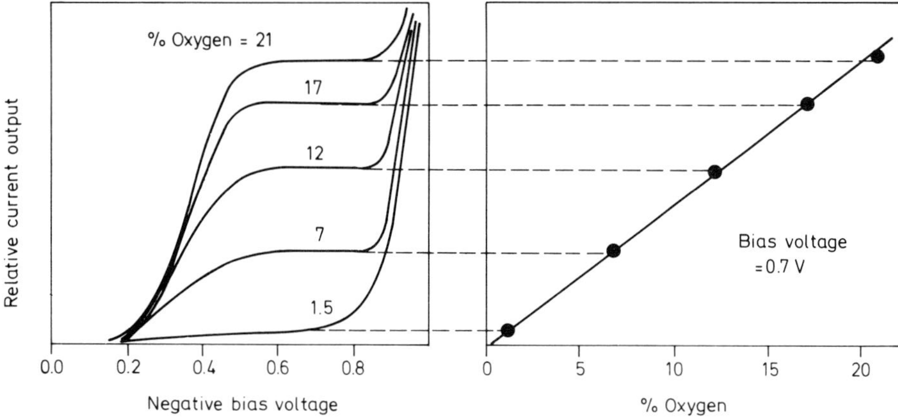

Fig. 1. Polarogram and calibration curve (Cobbold[34])

surface. When the negative bias voltage is further increased, the current output of the electrode increases rapidly due to other reactions, mainly, the reduction of water to hydrogen[37]. If a fixed voltage in the plateau region of the currentvoltage diagram is applied to the cathode, then the current output of the electrode can be calibrated to the dissolved oxygen (Fig. 1). It has to be noted that the current is proportional not to the actual concentration but to the activity or equivalent partial pressure of dissolved oxygen, which is often referred to as *"oxygen tension"*. A fixed voltage between −0.6 and −0.8 V is usually selected as the bias voltage (or polarization voltage) when using Ag/AgCl as the reference electrode[47].

When the cathode, anode, and the electrolyte are separated from the measuring medium with a plastic membrane, which is permeable to gas but not to most of the ions, and when most of the mass transfer resistance is confined in the membrane, the electrode system can measure oxygen tension in various liquids. This is the basic operating principle of the *membrane-covered polarographic DO probe* (Fig. 2a).

For polarographic electrodes, the reaction proceeds as follows[37]:

$$\begin{aligned}
&\text{Cathodic reaction: } && O_2 + 2\,H_2O + 2e^- && \rightarrow H_2O_2 + 2\,OH^- \\
& && H_2O_2 + 2e^- && \rightarrow 2\,OH^- \\
&\text{Anodic reaction: } && Ag + Cl^- && \rightarrow AgCl + e^- \\
&\text{Overall reaction: } && 4\,Ag + O_2 + 2\,H_2O + 4\,Cl^- && \rightarrow 4\,AgCl + 4\,OH^- .
\end{aligned}$$

The reaction tends to produce alkalinity in the medium together with a small amount of hydrogen peroxide[37]. Forbes and Lynn[50] postulated two principal pathways for the reduction of oxygen at the noble metal surface. One is a 4 electron pathway where the oxygen in the bulk diffuses to the surface of the cathode and is converted to H_2O via H_2O_2 (path a in Fig. 3). The other is a 2-electron pathway where the intermediate H_2O_2 diffuses directly out of the cathode surface into the bulk liquid (path b in Fig. 3). They stated that the oxygen reduction path changes depending on surface condition of

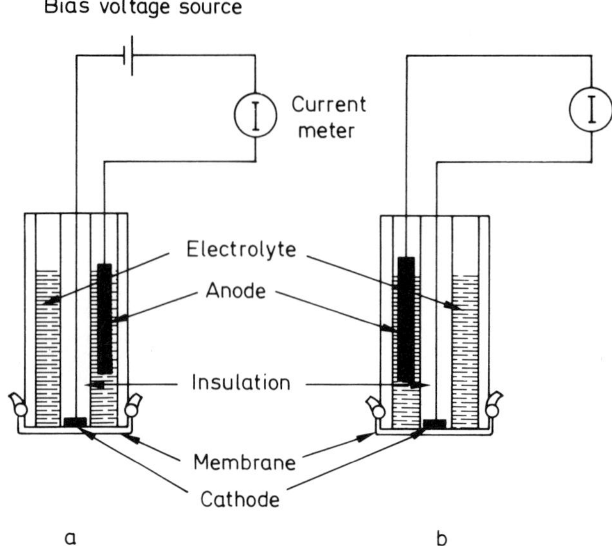

Fig. 2. Basic arrangements for (a) polarographic electrode and (b) galvanic electrode

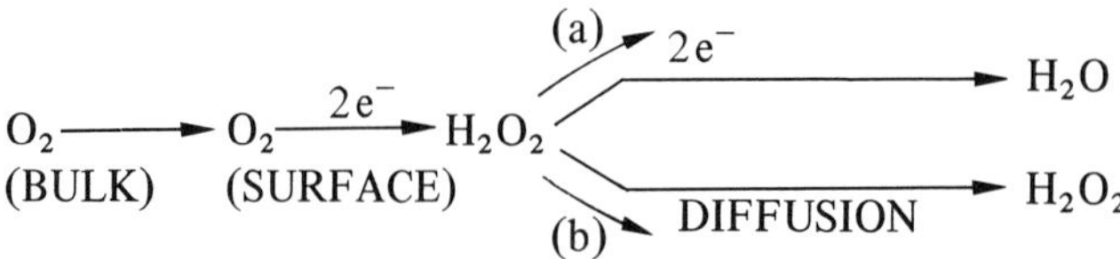

Fig. 3. Alternative pathways of oxygen reduction at the cathode surface (Forbes and Lynn[50])

the noble metal. This is probably the cause for time-dependent current drift of polarographic probes[9]. Since the hydroxyl ions are constantly being substituted for chloride ions as the reaction proceeds, KCl or NaCl has to be used as an electrolyte. When the electrolyte becomes depleted of Cl^-, it has to be replenished.

The *galvanic electrode* (Fig. 2b) is different from the polarographic type in that it does not require external voltage source for the reduction of oxygen at the cathode. When a relatively basic metal such as zinc, lead or cadmium is used as the anode and a relatively noble metal such as silver or gold is used as the cathode, the voltage generated by the electrode pair is sufficient for a spontaneous reduction of oxygen at the cathode surface. The electrode reaction of the silver-lead galvanic probe is as follows[104, 106]:

Cathodic reaction: $O_2 + 2\,H_2O + 4e^- \rightarrow 4\,OH^-$

Anodic reaction: $Pb \rightarrow Pb^{2+} + 2e^-$

Overall reaction: $O_2 + 2\,Pb + 2\,H_2O \rightarrow 2\,Pb(OH)_2$.

As shown above, the oxygen is reduced via four-electron reaction. Unlike the polarographic probe, the electrolyte does not participate in the reaction but the anode surface is gradually oxidized. Therefore, the probe life depends on the available surface area of the anode.

Whether the polarization voltage is applied internally (galvanic) or externally (polarographic), the operating principle of the electrode remains the same. For both types of probe, interference on measurement is expected when gases that reduce at 0.6–1.0 V are present in the test medium. Examples are halogens (Cl_2, Br_2, I_2) and oxides of nitrogen[13)]. Hitchman[64)] described in detail the electrochemistry of oxygen reduction.

3.2 Theory of Operation

The basic principle of measurement for the membrane-covered DO probes can be summarized as follows: provided that the oxygen diffusion is controlled by the membrane covering the cathode, the current output of the probe is proportional to the oxygen activity or the partial pressure in the liquid medium. The behavior of the probe can be predicted by using a simplified electrode model. For a mathematical analysis, the following assumptions are made:

1. The cathode is well polished and the membrane is tightly fit over the cathode surface so that the thickness of electrolyte layer between the membrane and the cathode is negligible.
2. The liquid around the probe is well agitated so that the partial pressure of oxygen at the membrane surface is the same as that of the bulk liquid.
3. Oxygen diffusion occurs only in one direction, perpendicular to the cathode surface.

This is the so-called one layer model[2, 7, 90)] but it can be extended to include the effects of other layers as will be shown later.

Suppose the electrode is immersed in a well-agitated liquid and, at time zero, the oxygen partial pressure of the liquid is changed from zero to p_0. According to Fick's 2nd law, the unsteady-state diffusion in the membrane is described as follows:

$$\frac{\partial p}{\partial t} = D_m \frac{\partial^2 p}{\partial x^2}, \tag{1}$$

where D_m is the oxygen diffusivity in the membrane and x is the distance from the cathode surface (Fig. 4a). The initial and boundary conditions are:

$$p = 0 \qquad \text{at } t = 0, \tag{2}$$

$$p = 0 \qquad \text{at } x = 0, \tag{3}$$

$$p = p_0 \qquad \text{at } x = d_m, \tag{4}$$

where d_m is the membrane thickness. The first boundary condition [Eq. (3)] assumes very fast reaction at the cathode surface. This condition was experimentally verified by Baumgärtl et al.[12)].

The solution of Eq. (1) with the boundary conditions is (2):

$$\frac{p}{p_0} = \frac{x}{d_m} + \sum_{n=1}^{\infty} \frac{2}{n\pi} (-1)^n \sin \frac{n\pi x}{d_m} \exp(-n^2 \pi^2 D_m t/d_m^2). \tag{5}$$

The current output of the electrode is proportional to the oxygen flux at the cathode surface[82)]:

$$I = NFAD_m \left(\frac{\partial c}{\partial x}\right)_{x=0} = NFAP_m \left(\frac{\partial p}{\partial x}\right)_{x=0} , \tag{6}$$

where N, F, A, and P_m are the number of electrons per mole of oxygen reduced, Faraday's constant, surface area of the cathode, and oxygen permeability of the membrane, respectively. The permeability, P_m, is related to the diffusivity, D_m, by

$$P_m = D_m S_m , \tag{7}$$

where S_m is the oxygen solubility of the membrane. From Eqs. (5) and (6), the current output of the electrode as a function of time, I_t is derived as follows[2,17,90]:

$$I_t = NFA(P_m/d_m)p_0 \left[1 + 2 \sum_{n=1}^{\infty} (-1)^n \exp(-n^2 \pi^2 D_m t/d_m^2)\right] . \tag{8}$$

The pressure profile and the current output under steady-state conditions can be obtained from Eqs. (5) and (8), respectively:

$$\frac{p}{p_0} = \frac{x}{d_m} \tag{9}$$

and

$$I_s = NFA(P_m/d_m)p_0 . \tag{10}$$

At steady-state, the pressure profile in the membrane is linear (Fig. 4a) and the electrode current is proportional to the oxygen partial pressure of the bulk liquid. Equation (10) forms the basis for DO probe measurements.

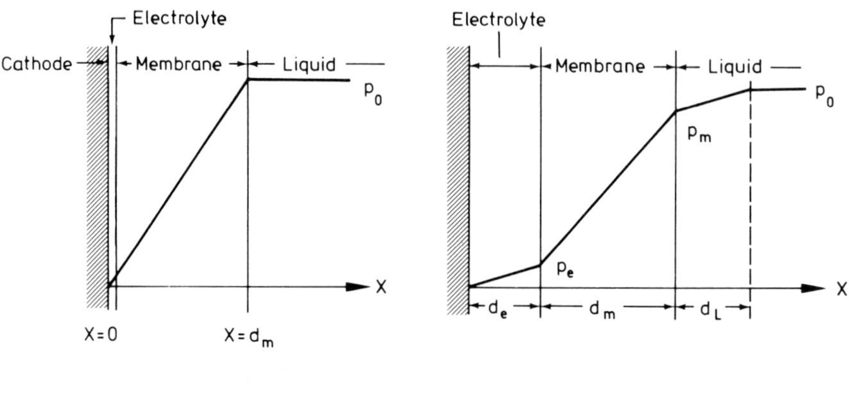

Fig. 4. (a) One-layer electrode model (b) three-layer electrode model

Another important consideration is the time response of the probe. According to Eq. (8), the probe response depends on the probe constant, k, defined as follows[15, 90]:

$$k = \frac{\pi^2 D_m}{d_m^2} . \tag{11}$$

A large k, which means a thin membrane and/or a high D_m, results in a fast probe response. However, these conditions tend to weaken the assumption of membrane-controlled diffusion. Thus, a compromise has to be made for an optimum probe performance.

In reality, the assumptions 1 and 2 made earlier are not entirely satisfactory. Often, there exists a finite thickness of electrolyte layer between the cathode and the membrane because of the roughness of the cathode surface. Also, a stagnant liquid film always exists outside the membrane even at very high liquid velocity. A more realistic model of the electrode has to be the one shown in Fig. 4b, where all three layers, namely, the electrolyte, the membrane and the liquid film are considered. Several authors[15, 17, 106] used two layer models incorporating the electrolyte layer and the membrane, while others[92, 128] discussed the effect of liquid film.

The effect of different layers on electrode behavior can be estimated by using the "one layer" model. At steady state, the oxygen flux, J, through each layer in Fig. 4b becomes identical:

$$\begin{aligned} J &= K_0 p_0 \\ &= k_{LM}(p_0 - p_m) \\ &= k_m(p_m - p_e) \\ &= k_e p_e , \end{aligned} \tag{12}$$

where K_0 is the overall mass transfer coefficient and small k's represent individual mass transfer coefficients corresponding to the liquid film (k_{LM}), the membrane (k_m) and the electrolyte (k_e), respectively. The overall mass transfer resistance, $1/K_0$, is then expressed as the sum of the individual resistances:

$$\frac{1}{K_0} = \frac{1}{k_{LM}} + \frac{1}{k_m} + \frac{1}{k_e} . \tag{13}$$

Equation (13) can be rewritten by using the oxygen permeability and the thickness of each layer:

$$\frac{1}{K_0} = \frac{d_L}{P_L} + \frac{d_m}{P_m} + \frac{d_e}{P_e} , \tag{14}$$

where d_L, d_e, P_L, and P_e are liquid film thickness, the electrolyte thickness, the oxygen permeability of the liquid film and that of the electrolyte layer, respectively. A completely stagnant liquid film was assumed here, although it is more accurate to use the convective mass transfer coefficient, k_{LM}, directly[92].

The condition for a membrane-controlled diffusion becomes:

$$\frac{d_m}{P_m} \gg \frac{d_L}{P_L} + \frac{d_e}{P_e} . \tag{15}$$

This means that a relatively thick membrane with a low oxygen permeability is required, which contradicts the requirement for a fast probe response. For a given cathode geometry, the resistance of the electrolyte is more or less fixed. Also, since the electrolyte is contained inside the membrane, it does not affect the measurement. Therefore, the condition for accurate measurements of dissolved oxygen becomes:

$$\frac{d_m}{P_m} + \frac{d_e}{P_e} \gg \frac{d_L}{P_L} . \tag{16}$$

When the individual resistances are taken into account, the steady state current output can be written as,

$$I_s = NFA(P_m/\bar{d})p_0 , \tag{17}$$

where $\bar{d}$ is defined as,

$$\bar{d} = d_m + \frac{P_m}{P_L} d_L + \frac{P_m}{P_e} d_e . \tag{18}$$

In this case, the probe constant, k, is modified as follows:

$$k = \frac{\pi^2 D_m}{\bar{d}_t^2} , \tag{19}$$

where

$$\bar{d}_t = d_m + \sqrt{\frac{D_m}{D_L}} d_L + \sqrt{\frac{D_m}{D_e}} d_e . \tag{20}$$

Equations (17) and (19) show that the steady-state current decreases and the probe response time increases when there is a significant mass transfer resistance in the liquid film around the membrane. Normally, probes are operated such that the effect of liquid film resistance is negligible: this is achieved by using membranes of low oxygen permeability and by a vigorous agitation of the liquid around the probe.

Sometimes, the assumption of one dimensional diffusion (assumption 3) is not satisfactory, especially when the cathode diameter is small compared to the membrane thickness. Often, the probe response shows hysteresis[15, 61, 79)] and doubling of the cathode area does not result in doubling of the output current[131)]. Electrode models incorporating lateral or sideways diffusion[69, 80)] and multi-layer, multi-region models[93)] are helpful for understanding the electrode behavior in these situations.

3.3 Oxygen Microelectrodes

The measurement of local oxygen tension in microbial pellets and films is often necessary to understand the mass transfer mechanisms involved[28, 66)]. The capability of measuring local oxygen concentration inside the liquid diffusion boundary layer enhanced our understanding of the oxygen transport process[27, 86, 87, 139)]. Oxygen microelectrodes have been used for these purposes, which have long been applied in physiology for tissue oxygen measurements[136)].

Figure 5 shows a basic arrangement for oxygen microelectrode measurements. This is similar to the membrane-covered polarographic probe described earlier except that the cathode and the reference electrode (anode) are used separately. Thus, an electrolyte is required in the medium for dissolved oxygen measurements. The analysis of the membrane-covered microcathode is similar to that

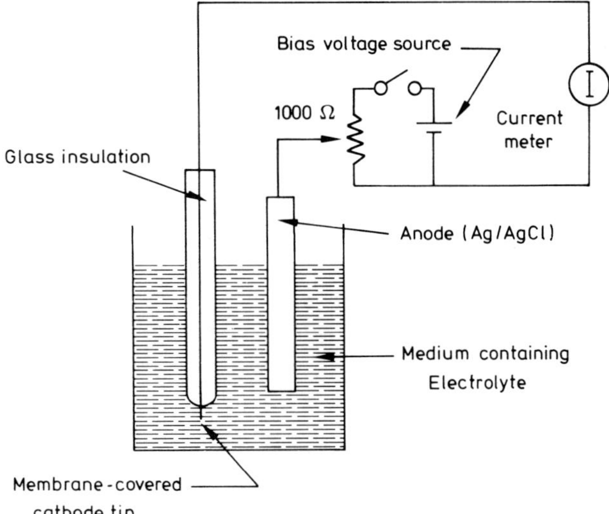

Fig. 5. Basic arrangement for oxygen microelectrode

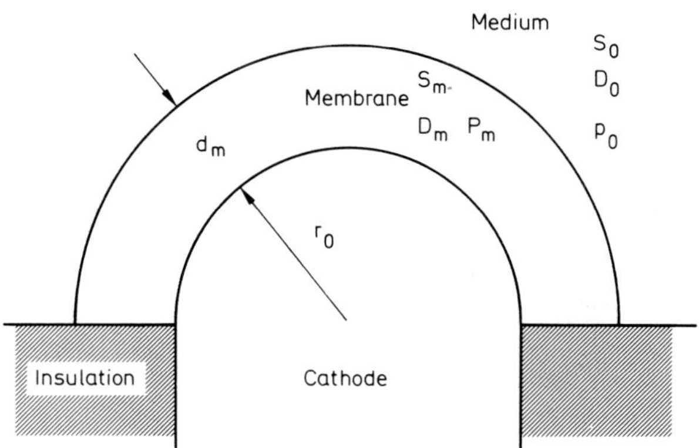

Fig. 6. Model of microcathode

of the macroprobes given earlier. Since the shape of the cathode tip is normally conical[86)], a half spherical cathode with a radius of r_0 is used as a model. Suppose the cathode is immersed in an infinite medium whose undisturbed oxygen partial pressure is p_0 (Fig. 6). Considering the symmetry of the half-sphere, the governing diffusion equations with boundary conditions are:

$$\frac{\partial p}{\partial t} = D_m \left(\frac{\partial^2 p}{\partial r^2} + \frac{2}{r} \frac{\partial p}{\partial r} \right) \quad \text{for} \quad r_0 \leqslant r \leqslant r_0 + d_m \ , \tag{21}$$

$$\frac{\partial p}{\partial t} = D_0 \left(\frac{\partial^2 p}{\partial r^2} + \frac{2}{r} \frac{\partial p}{\partial r} \right) \quad \text{for} \quad r \geqslant r_0 + d_m \ , \tag{22}$$

$$p = 0 \quad \text{at} \quad r = r_0 \ , \tag{23}$$

$$p = p_0 \quad \text{at} \quad t = 0 \ , \tag{24}$$

$$p = p_0 \quad \text{at} \quad r = \infty \ , \tag{25}$$

$$-P_0\left(\frac{\partial p}{\partial r}\right)_{r=(r_0+d_m)^+} = -P_m\left(\frac{\partial p}{\partial r}\right)_{r=(r_0+d_m)^-} . \tag{26}$$

The analytical solutions to these equations[56)] describe the transient behavior of the electrode. Steady-state solutions are:

$$p = p_0\left(\frac{r_0+d_m}{d_m}\right)\left(\frac{s}{1+s}\right)\left(1-\frac{r_0}{r}\right) \quad \text{for} \quad r_0 \leqslant r \leqslant r_0 + d_m , \tag{27}$$

$$p = p_0\left[1-\left(\frac{1}{1+s}\right)\left(\frac{r_0+d_m}{d_m}\right)\right] \quad \text{for} \quad r \geqslant r_0 + d_m , \tag{28}$$

where

$$s = \frac{P_0\, d_m}{P_m\, r_0} . \tag{29}$$

The steady state current output can be obtained from Eq. (6):

$$I_s = 2\,NFP_m r_0\left(\frac{r_0+d_m}{d_m}\right)\left(\frac{s}{1+s}\right)p_0 . \tag{30}$$

Equations (27) and (28) are plotted in Fig. 7 for different P_0/P_m when the membrane thickness is twice the diameter of the cathode. It is shown that the pressure gradient becomes confined within the membrane with increasing P_0/P_m. The condition for 99% of the pressure gradient to be confined inside the membrane can be obtained from Eq. (27):

$$\frac{P_0\, d_m}{P_m\, r_0} \geqslant 99 . \tag{31}$$

When this condition is satisfied, the local oxygen tension can be measured with a high spacial resolution and, for liquid phase measurements, the probe output is not affected by liquid velocity. In this case, the steady state current becomes independent of the oxygen permeability of the medium:

$$I_s = 2\,NFP_m r_0\left(\frac{r_0+d_m}{d_m}\right)p_0 . \tag{32}$$

This equation shows that the current is proportional to the oxygen partial pressure of the medium but it is not exactly proportional to the surface area of the cathode.

For a cathode with membrane-controlled diffusion, the 95% response time of the probe, $\Gamma_{95\%}$, is given approximately as follows[57)]:

$$\Gamma_{95\%} = \frac{d_m^2}{2\,D_m} . \tag{33}$$

Equation (33) shows that the response time of the probe is inversely proportional to the probe constant, k, defined by Eq. (11). The probes in this category have cathode diameters of around 1 μm and, with a proper membrane, $\Gamma_{95\%}$ is much less than 1 s[11, 18, 87, 134, 149)].

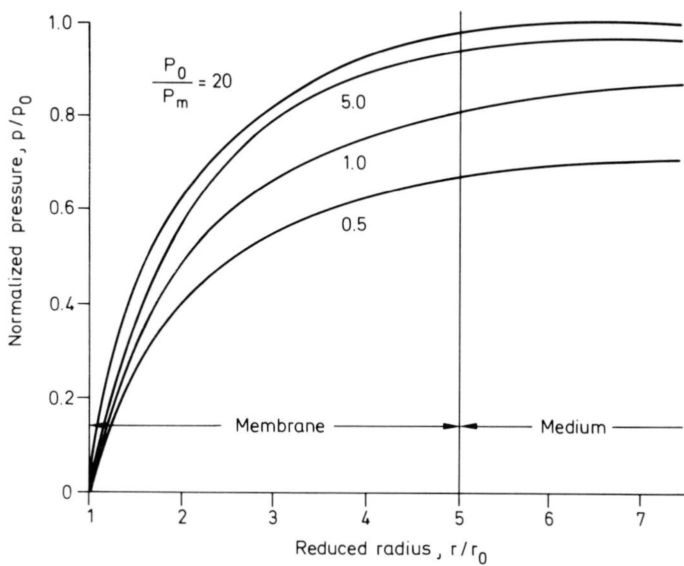

Fig. 7. Effect of (P_o/P_m) on steady-state pressure profile (Lee et al.[87])

4 Design of Electrodes

Although the membrane-covered DO electrode basically consists of a cathode, an anode, and the electrolyte, numerous designs of the probe have appeared in the literature during the past 20 years[47]. In designing DO probes, the following requirements are generally considered:

the calibration has to be stable over a long period;
the current output of the probe has to be sufficiently large and linear with DO;
the effect of liquid flow on probe performance has to be small;
the response time has to be fast;
the measurement has to be independent of temperature change of the medium;
and the probe must withstand high pressure and repeated autoclaving.

In actual probe design, some of the above requirements are emphasized more than others depending on specific applications.

Over 20 companies in the U.S. offer different types of DO probe for different applications. Some of them are listed in Table 1. Fatt[47] gave a comprehensive review of various probe designs. Shown in Table 2 are comparisons of six representative probes of the original design. Details of construction methods, selection of electrode components and design considerations are given below.

Table 1. Some oxygen electrode manufacturers

Type	Maker	Address	Cathode	Anode	Remark
Polarographic	Beckman Instrument	Irvine, California	Pt	Ag	Laboratory and field use
	Delta Scientific	Lindenhurse, New York	Au	Ag	Field use
	Instrumentation Laboratory	Lexington, Massachusetts	Pt	Ag	Steam-sterilizable probes
	Orbisphere	York, Maine	Au	Ag	Low drift, low DO; laboratory and field use
	Yellow Springs Instrument	Yellow Springs, Ohio	Au	Ag	Laboratory and field use
Galvanic	Biomarine Industries	Malvern, Pennsylvania	Au	Pb	Hospital and industrial use
	Electronic Instrument Ltd.	Surrey, England	Ag	Pb	Mackereth type
	New Brunswick Scientific	Edison, New Jersey	Ag	Pb	Borkowski-Johnson type, steam-sterilizable
	Precision Scientific	Chicago, Illinois	Ag	Pb	Mancy type; field use
	Rexnord Instrument	Malvern, Pennsylvania	Pt	Pb	Low DO; boiler feedwater applications
Microprobe	Transidyne General	Ann Arbor, Michigan	Pt	Ag	Local oxygen measurement

Table 2. Comparison between different types of DO electrodes

Electrode type	Clark	Mancy	Mackereth	Borkowski-Johnson	Kimmich-Kruezer	Microelectrode
Principle	Polarographic	Galvanic	Galvanic	Galvanic	Polarographic	Polarographic
Electrode metals (cathode-anode)	Pt–Ag/AgCl	Ag–Pb	Ag–Pb	Ag–Pb	Pt–Ag/AgCl	Pt–Ag/AgCl
Cathode: shape	Disk	Disk	Tubing	Spiral, disk	Thin band ring	Half-spherical
size	2 mm dia.	6 mm dia.	3.6 cm dia. 5.8 cm long	6 mm dia.	3 μm width 1 mm dia. ring	0.2–1.0 μm dia.
Membrane (thickness)	Polyethylene 25 μm	Polyethylene 25 μm	Polyethylene 75 μm	Teflon, FEP 50 μm	Teflon, FEP 6 μm	Polystyrene 1 μm
Electrolyte	sat. KCl	1 M KOH	sat. K_2HPO_4	pH 3, acetate buffer	pH 8, phosphate buffer	0.2 M KCl
Current output: (25 °C), μA						
Air saturation	6	10	200	10	0.12	$3(10^{-4})$
Zero-oxygen	0.001–0.01	0.2	negligible	0.001	0.0005	$1(10^{-5})$
Response time (25 °C)	15 s (95%)	5.6 s (95%)	60 s (90%)	60 s (95%)	0.4 s (95%)	0.1–0.05 s (95%)
Stability	2% drift/day	No loss of sensitivity for 3 weeks	No change in calibration for 6 months	0.5% drift/day 1 year life	2% drift/day	1–2% drift/hour
Ref.	31, 131	105	104	21, 71	75	11, 86

4.1 Construction Methods

4.1.1 Clark-type Electrode

As shown in Fig. 8a, this probe is characterized by a flat disk type cathode and a pool of electrolyte in which a reference electrode (Ag/AgCl) is immersed. Although the size of the cathode, the membrane material and the electrolyte differ widely, this design is most popular in commercial DO probes for use in the laboratory or in the field.

Clark-type electrodes often show current drift during initial "break-in" periods. After that, the current output remains relatively stable although the response and the calibration may change with use[15, 67, 80, 88]. The probe malfunction is caused by AgCl deposition on the anode surface[88], a deposition of silver on the cathode[131], a depletion of Cl^- from the electrolyte, or a loose membrane. However, with proper cleanings of the electrode, membrane replacements, electrolyte replenishments and frequent calibrations, these probes can be used for a long period.

With a 25 μm Teflon membrane, 95% response time of 15–20 s is usually obtained. But these sensors show a response hysteresis: the response to an increased oxygen tension is faster than that to a decreased oxygen tension. This phenomenon is caused by the electrolyte acting as a reservoir of dissolved oxygen[80] and/or the accumulation and slow decomposition of hydrogen peroxide in the vicinity of the cathode[99].

Some of the well-designed Clark-type probes showed very stable calibration and extremely low residual current. Orbisphere polarographic probe[117], which uses gold as the cathode, showed a drift of less than 0.01 ppm during 60 days of continuous measurement. Orbisphere (York, Maine) also manufactures a polarographic probe capable of measuring parts per billion range of DO. The improved performance of the probe came from careful choice of materials, good mechanical design and the use of extremely stable amplifier circuit.

4.1.2 Mancy Electrode

The galvanic probe originally designed by Mancy et al.[106] is shown in Fig. 8b. The major difference from the Clark electrode is the elimination of the electrolyte chamber. Instead, a thin film of electrolyte is placed between the cathode and the membrane. This is probably the reason why this electrode did not show the response hysteresis. Due to a relatively large diameter (0.6 cm) cathode employed, a microammeter could be directly connected to the probe. However, the probe showed very high flow dependency of the output current: stirred/unstirred current ratio of 20 was reported[105].

An improved probe stability was reported compared with earlier polarographic probes but the useful probe life may be somewhat restricted because the available surface area of the anode is relatively small. The anode surface is gradually oxidized with use until the probe cease to function.

4.1.3 Mackereth Electrode

Noting that the earlier Clark electrode and the Mancy electrode lacked long term stability and produced small currents (on the order of μA), Mackereth[104] designed a

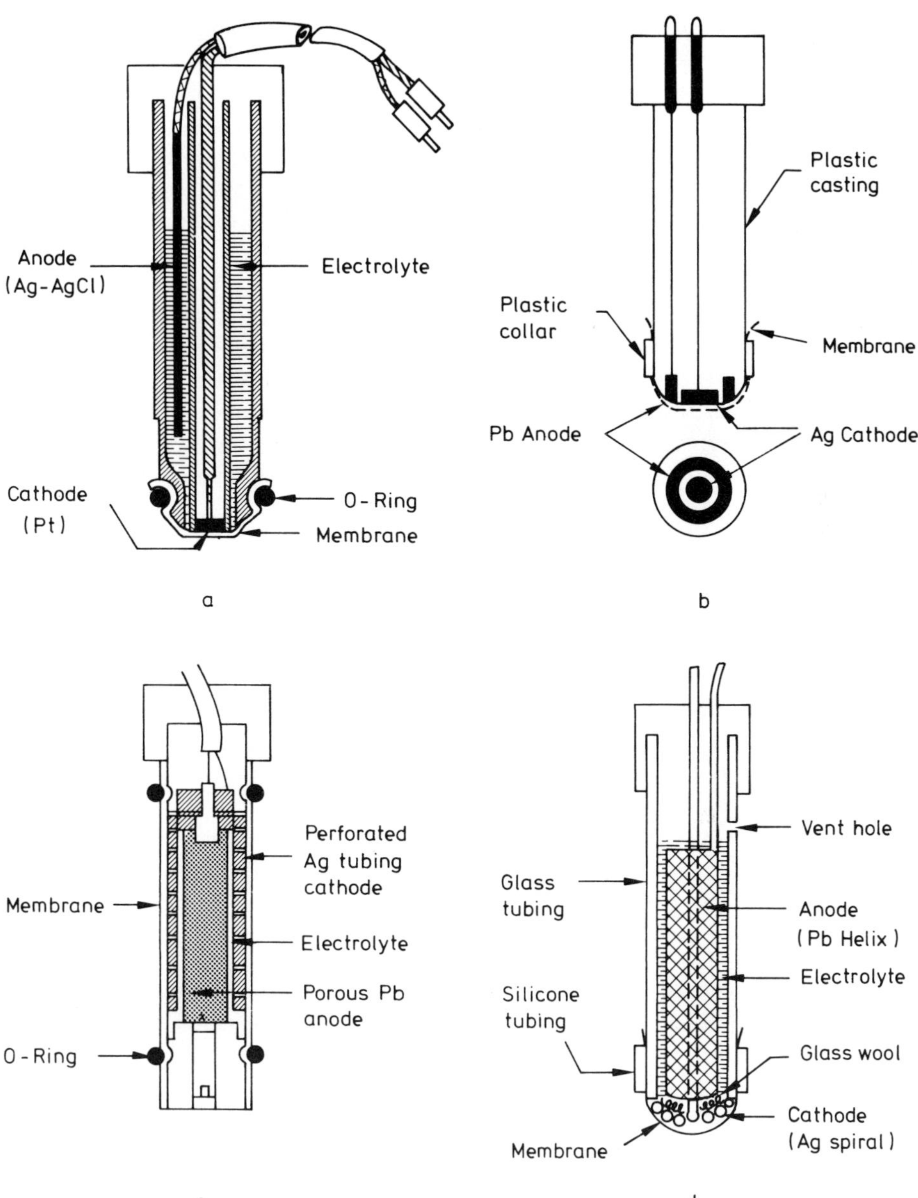

Fig. 8. Construction of various electrodes: (a) Clark[31)], (b) Mancy[105)], (c) Mackereth[104)], (d) Borkowski-Johnson[21)]

probe to solve these problems. In his design, a perforated silver tubing was used as the cathode and a massive shot of porous lead was used as the anode (Fig. 8c). The current output was much higher than those of other probes so that an ordinary current meter can be directly connected to the probe without amplification. The sensitivity was stable over many months of continuous operation. With a 25 μm Teflon FEP membrane, this probe gave a 90% response time of about 1 min. Harrison[61)] reported hysteresis in the probe response due to the electrolyte solution acting as a reservoir for dissolved oxygen. This observation is similar to that of the Clark electrode described earlier.

Because of its long term stability, this probe has been used for monitoring the dissolved oxygen tension in continuous cultivation which lasts several weeks[61, 107, 126)].

When vertically inserted in an air-sparged cylindrical vessel, the Mackereth probe exhibited a "gross" response free from any effects of air bubbles ascending through the vessel, whereas other probes with cathodes at the tip showed interference due to bubbles touching the cathode[3)]. Since the cathode area is relatively large, a vigorous agitation of the liquid is required for a reliable operation and the oxygen consumption by the probe must be considered when using this probe for measuring oxygen in a small liquid volume. The probe may not be suitable for a viscous medium. The need for a tubing membrane makes it difficult to fabricate the probe in the laboratory.

4.1.4 Borkowski-Johnson Electrode

The original design was by Johnson et al.[71)] but it was improved later by Borkowski and Johnson[21)] for a longer life and a better stability. As shown in Fig. 8d the cathode is made from a silver spiral and a flattened lead wire forms the anode. A low pH acetate buffer is used as the electrolyte to prevent interference by dissolved CO_2. This probe has been used widely in biochemical engineering applications[42, 61, 120)]. The probe can withstand repeated steam sterilizations, is capable of operating for several months and has a linear response from below 0.02 to 150 mmHg of oxygen[21)].

With a 50 μm Teflon membrane, 90% response time of 1 min was reported[71)]. The response time varies depending on the direction of step change in oxygen tension. The upstep response is faster than the downstep response and the latter depends on the length of time the probe was exposed prior to the downstep[79)]. A vigorous agitation of the liquid (at least 60 cm s^{-1} for water) is required for reliable measurements[42)]. The probe may not be suitable for viscous liquids unless a thicker membrane is used.

4.1.5 Electrodes with Low Flow Dependency and Fast Response

The requirements of the probe for low flow dependency and fast response time oppose each other: a thicker membrane, needed for low flow dependency of probe sensitivity, gives slow response time. One solution to this problem is to use cathode with small diameters. For example, a 12 μm diameter platinum cathode covered with 6 μm Teflon membrane gives 99% response time of 1.2 s and requires very low flow velocity for proper operation[131)]. The problems are that the current output is very low (on the nA range) and often the stability of the probe is poor[73)].

Another approach is to use various shapes of narrow band cathodes. Locally, the narrowness of a thin band allows the advantages of small cathodes regarding the flow dependency and fast response, but since the total surface area of the cathode is relatively large, the probe gives large output current (on the μA range) and a better stability. Figure 9 shows various shapes of thin band cathodes. Usually, metal foils are imbedded in a casting epoxy and ground flat to produce the desired form. Figure 9 a shows a ribbon type cathode employed by a Beckman probe[13)], which requires only 5 cm s^{-1} of water velocity with a 25 μm Teflon membrane, compared with 30–70 cm s^{-1} for other probes. Kimmich and Kreuzer[75)] used a ring type cathode (Fig. 9 b) which showed a 95% response time of 0.4 s and a flow requirement of 5 cm s^{-1} in blood with a 6 μm Teflon membrane. Jones et al.[72)] further increased the cathode surface by using an S-shaped cathode in their disposable galvanic sensor (Fig. 9 c) for use in deep water.

A multicathode approach was used with some degree of success[48, 137)]. Fatt and St. Helen[48)] used 18 cathodes, each 25 μm in diameter, in a single probe body which was able to measure oxygen tension in the range of 0–1 mmHg with a linearity better than 1%. Siu and Cobbold[137)] described a multicathode Clark-type probe that is fabricated by using integrated circuit technique. They were able to lay down 161 gold cathodes each 7 μm diameter and spaced 60 μm apart to give a sensor with a diameter of about 0.6 mm. The probe showed extremely low residual current, 0.8% linearity in the range of 0–760 mm Hg of oxygen partial pressure, and very low flow dependency of probe sensitivity (flow increased the output 1% above the static value).

One of the *steam sterilizable probes* of Instrumentation Laboratory (Lexington, Massachusetts) employed a double membrane, which showed both low flow sensitivity and a reasonably fast response time. The inner, current-determining membrane was a 25 μm Teflon membrane and the outer membrane was a highly permeable, 150 μm silicone membrane[26)]. This probe, which had a 250 μm diameter platinum cathode, showed only 2% change in current between an agitated and non-agitated solution, and 98% response in 50 s.

Since the condition of membrane-controlled diffusion is readily obtainable with probes in this category, they can be used in viscous liquids with better accuracy compared with other probes.

4.1.6 Oxygen Microelectrodes

The major considerations with these probes are: a high spacial resolution and a fast response time. When the cathode diameter becomes less than 1 μm, even a bare metal cathode (Fig. 10 a) becomes insensitive to liquid flow and measures local oxygen tension[136)]. The probe performance is improved by covering the cathode with a membrane (Fig. 10 b)

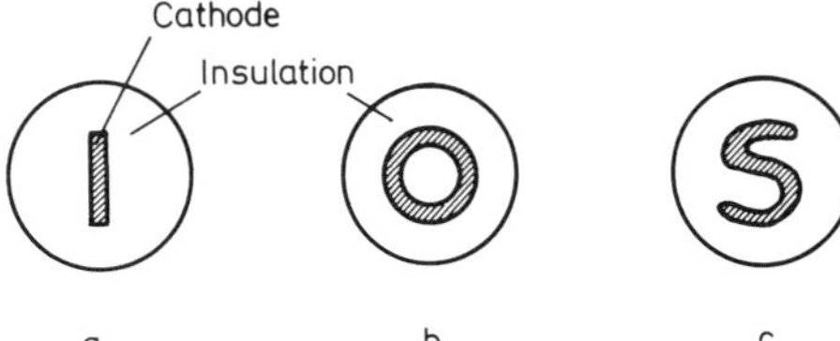

Fig. 9. Different shapes of cathode: (a) ribbon, (b) ring, (c) S-shape

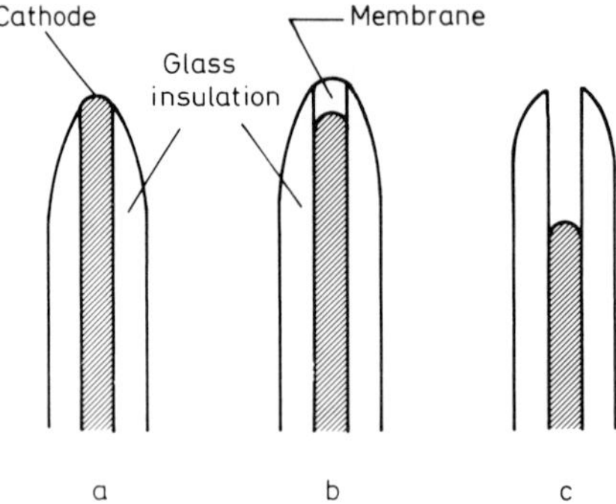

Fig. 10. Different shapes of microcathodes: (a) bare, (b) membrane-covered, (c) recessed (Lee[86])

or by extending the insulation (Fig. 10c) such that the diffusion gradient is confined inside the recess[148].

Details of construction of these probes were given elsewhere[18, 86, 134, 148]. Usually, a thick platinum wire is first etched in an electrolyte solution to a fine point of 0.2–1 μm diameter and then insulated with a thin layer of glass[86]. The membrane is applied by a dip coating. Polystyrene is most popular[136] but other materials have been used. The membrane coating has to be done carefully[129] to obtain optimum result.

Although the capability of local oxygen measurement is attractive, several problems exist with these microprobes: they are relatively difficult to make, requiring practice and experience; extremely fragile; unstable; and the useful probe life is short. Since the current output of the probe is extremely low (on the order of 10^{-10} to 10^{-11} A in air-saturated water), a special amplifier is required together with a careful grounding technique to prevent noise[86].

4.2 Electrode Metals

Sawyer and Interrante[127] studied the reduction of dissolved oxygen at Pt, Pd, Ag, Ni, Au, Pb, and other metal electrodes. They found that oxygen reduction at pre-oxidized metal electrodes is pH-dependent, whereas for pre-reduced metal electrodes, the reduction was pH-independent. They also showed that the reaction mechanism of oxygen reduction is complex due to an oxide film coating on metal surfaces except for Ag and Au. Evans and Lingane[44] reported that even the Au surface showed aging effects. Mancy et al.[106] preferred Ag for a better stability, but Ag is readily poisoned by sulfur impurities[64].

For polarographic probes, platinum[31], gold[76], silver[118], or rhodium[13] have been used as the cathode and Ag[118], Ag/AgCl[31], Ag/Ag_2O[30], or calomel[69] as the reference electrode. Gold is generally preferred to platinum as the cathode material, because it is less susceptible to poisoning by noxious gas, notably, H_2S[64]; the reaction at the cathode surface is less complicated[44, 127]; and surface aging is less pronounced[64]. However, gold may not be convenient for applications in steam-sterilizable probes and in microelectrodes since gold and glass cannot be fused together. Gold-plating method has been used in making microelectrodes to circumvent this problem[148].

The reference electrode has to maintain a stable reference voltage for a good performance of the polarographic probe. Also, it has to have a large surface area to avoid polarization[37)]. Ag/AgCl is normally used as the reference electrode. However, Carrit and Kanwisher[30)] preferred Ag/Ag_2O because it gave better stability for their probe. When the chloride containing electrolyte is used, the chloride ion concentration will fall as it is consumed by the anode reaction and will be replaced by OH^- generated by the oxygen reduction at the cathode. Consequently, Ag/AgCl will gradually change to Ag_2O, and since the reference voltage are different (+0.222 V for Ag/AgCl and +0.35 V for Ag/Ag_2O), this may cause a change in probe sensitivity unless the probe has a wide current-voltage plateau[37)].

The *aging effect* of polarographic probes such as the change in calibration or unstable probe sensitivity is attributed to the deposition of silver ions on platinum[131)], the oxidation of the catalytic surface[44)] or an excessive deposition of AgCl on the reference electrode[88)]. The cathode surface can be mechanically cleaned with soft scouring powders and a wet leather[131)] or by tooth paste[47)]. Excessive AgCl deposits can be removed by washing with 15% NH_4OH[88)]. These procedures normally rejuvenate an aged probe. Kessler[73)] reported unstable behavior of platinum microcathodes, but Barr et al.[9)] showed that periodical anodization of Pt improved the stability.

For galvanic probes, silver as the cathode and lead as the anode are most common[24, 104, 106, 118)] but silver-aluminum[3)], platinum-aluminum[54)], platinum-lead[43, 123)], gold-zinc[70)], and gold-lead[19)] pairs have also been used. Although the galvanic probes suffer less from poisoning and survive autoclaving with greater reliability[142)], the probe life is limited by the available surface area of the anode due to gradual oxidation of anode surface. The useful life-time of a given probe depends on the current drain just like an electrical battery. In other words, a probe can be used much longer when it is used for monitoring low, rather than high oxygen tension. Dead probes can be rejuvenated by dissolving away the oxide layer on the anode surface with 20% CH_3COOH[104)] or 20% HCl[126)].

4.3 Electrolytes

Since the electrode reaction occurs in the electrolyte solution, the composition, pH, and the volume the electrolyte are directly related with probe stability. For polarographic probes, the electrolyte takes part in the reaction (depletion of Cl^-: see Sect. 3.1), so a refill of electrolyte is necessary in regular intervals. In general, pH of the electrolyte does not affect oxygen reduction on a clean metal surface but an adverse effect was observed for an oxidized metal cathode[127)]. The solubility of the electrode metals in the electrolyte solution has to be low for probe stability. Sometimes, the polarization voltage may change depending on electrolyte concentration[29)]. In this case, the change in electrolyte concentration due to evaporation or diffusion of water through the membrane affects probe stability.

Table 3 shows various electrolyte solutions used by different investigators. For polarographic probes, KCl is the most common electrolyte, which gives a constant calibration well over 48 h[100)]. Sometimes, CO_2 present in the medium, such as blood or cultivation media, permeates through the membrane and alters pH of the electrolyte solution. To eliminate this small CO_2 effect, Severinghaus[131)] used 0.5 M $NaHCO_3$ + 0.1 M KCl,

Table 3. Various electrolytes used in DO probes

		Electrolyte	Ref.
Polarographic	Clark	sat. KCl	31
	Lübbers et al.	0.2 M KCl	101
	Severinghaus	0.5 M $NaHCO_3$ + 0.1 M KCl	131
	Carrit and Kanwisher	0.5 M KOH	30
	Kimmich and Kreuzer	Phosphate buffer ar pH 8–9	75
	Pittman	NH_4Cl	119
Galvanic	Mancy et al.	1 M KOH	105
	Mackereth	sat. $KHCO_3$	104
	Rowley	sat. $NaHCO_3$	126
	Parker and Clifton	10% by wt. K_2HPO_4	121
	Borkowski and Johnson	acetate buffer	21
	Brookman	1 M K_2HPO_4	24
	Evangelista et al.	50% by wt. KI	43
	Harrison and Melbourne	30% by wt. K_2CO_3 + 10% by wt. $KHCO_3$	59

which had pH of 9. Buffered phosphate[75)] has also been used for the same purpose. Carrit and Kanwisher[30)] used 0.5 M KOH in their probe (Pt–Ag/Ag_2O) because it gave a better stability compared with KCl electrolyte. Since the hydroxide is a reaction product formed at the cathode, the hydroxide added initially as the electrolyte is expected to make the electrode reaction to start under a condition which is similar to that obtainable after a long period of use. A problem with a strong alkaline electrolyte is the interference by CO_2 [64)]. Pittman[119)] used NH_4Cl in his steam-sterilizable polarographic probe because KCl caused an early failure of the silicone tubing used as the membrane.

For galvanic probes, Mancy et al.[105)] preferred 1 M KOH to KCl, because KCl gave high residual current and it did not maintain a clean anode surface. The solubility of Ag is reported to be high in KCl[24)]. In the original Mackereth galvanic cell[104)], saturated $KHCO_3$ was used, but a mixture of saturated K_2CO_3 and $KHCO_3$ was found to be better by others[59, 126)] in preventing the effect of CO_2 on probe stability. When bicarbonate solution is used as the electrolyte, a whitish deposit of basic lead carbonate builds up on the anode, in comparison with KOH which keeps a clean anode. Borkowski and Johnson[21)] employed 5 M acetic acid + 0.5 M sodium acetate + 0.1 M lead acetate for the electrolyte in their steam-sterilizable galvanic cell. This electrolyte has a low pH (pH of 3) and hence the calibration and the cell life are not affected by the CO_2 permeating through the membrane during the monitoring of DO in cultivation media[42)].

Brookman[24)] tested most of the electrolytes listed in Table 3 and concluded that only KCl and K_2HPO_4 gave a linear response up to a high oxygen tension and also were steam-sterilizable. Since KCl gave a high residual current, he recommended using 1 M K_2HPO_4 as the electrolyte. He obtained a linear response of the electrode to oxygen partial pressure from 1.5×10^{-2} to 10^3 mm Hg. Evangelista et al.[43)] used 50% by weight KI for the electrolyte in their galvanic probe which could measure parts per billion range of dissolved oxygen. Sawyer and Interrante[127)] showed that, although the oxygen reduction is independent of pH, it changes with iodide concentration. They reported that the reaction involved formation of PtI_2 film on the platinum surface.

A rather serious problem with the electrolyte is the *loss of solvent through the membrane* by diffusion or evaporation. Several methods have been used to minimize this effect. One method was by having a large electrolyte reservoir, which can effectively supply solvent to the electrolyte film where the solvent loss occurs. In this case, the path between the reservoir and the electrolyte film has to be as long as possible to minimize the residual current[83)]. Another method is to have the electrolyte in the form of gel or paste. An added advantage is that the residual current becomes smaller because the oxygen permeability is normally lower in more viscous media. A third method is to add a small amount of deliquescent salt to the electrolyte. Hitchman[64)] showed that the addition of 0.01 M KH_2PO_4, which is a deliquescent salt, to 2.33 M KCl electrolyte prevented loss of solvent and lengthened probe life considerably. When not in use, the probe has to be stored in a water-saturated atmosphere or in water to prevent evaporation loss of electrolyte solution.

A recent development has been the use of a *solid electrolyte* instead of the conventional liquid form. Niedrach and Stoddard[114)] described a probe using ion exchange membrane as the electrolyte. Certainly, this would lead to a more rugged construction of the probe.

4.4 Membrane

An ideal membrane for use in DO probes has to have a relatively low oxygen permeability and a high oxygen diffusivity. The permeability has to be low to ensure membrane control of oxygen diffusion [Eq. (15)], whereas a high diffusivity gives fast probe response [Eq. (11)]. The membrane has to be mechanically strong and chemically inert. Since the current output is directly related to the thickness and the oxygen permeability of the membrane [Eq. (10)], the probe sensitivity is directly affected by change in membrane properties. Membrane swelling or change in the apparent oxygen permeability results in change of probe sensitivity. Other important factors are CO_2 permeability and water permeability. The water permeability of the membrane has to be low to prevent loss of water from the electrolyte solution, which causes an increase in electrolyte concentration and early failure of the probe. Low CO_2 permeability of the membrane is also desirable for probes to be used in aerobic cultures and in blood.

Teflon[71)], polyethylene[31)], and polypropylene[131)] have been most popular as the membrane material but silicon[119)], polystyrene[135)], and mylar[41)] have also been used. The properties of various membranes are shown in Table 4. It has to be noted that data on gas permeability and diffusivity vary widely depending on the method of measurement[68)]. According to the data shown in Table 4, polypropylene is better than Teflon in several aspects: it has a lower oxygen permeability; a lower CO_2 permeability; and yet a higher oxygen diffusivity. Polypropylene was recommended by Severinghaus for physiological works[131)] and by Kinsey and Bottomley for bioreactor applications[76)]. However, Teflon seems to be more popular in steam-sterilizable probes due to its higher

Table 4. Properties of various membranes

Membrane	O_2		CO_2	H_2O		Heat Resistance °C
	P_m [a]	D_m [b]	P_m [a]	P_m [a] (vapor)	Absorption (24 h, %)	
Teflon FEP	4.4	1.07	9.9	29	0.01	227–274
Polypropylene	1.2	1.62	3.9	51	0.005	132–149
Polyethylene						
low density	3.0	–	16.0	95	0.01	82–93
med. density	2.4	–	10.4	51	0.01	104
hi. density	1.1	–	3.4	22	Nil	121
Polystyrene	2.1	–	5.3	624	0.06	80–96
Mylar [c]						
(polyester)	0.05	–	0.09	148	0.8	300
Silicone [c]	480		2,530	16,900	–	–

[a] 10^{-10} cc $s^{-1} cm^{-1} (cm\ Hg)^{-1}$
[b] 10^{-7} cm^2 s^{-1}
[c] From Ref. 131
[d] From Ref. 2
All others from Ref. 110

heat resistance. An added advantage of the Teflon is its extremely low water permeability. Polystyrene is popular for microprobes because it sticks well on the glass insulation and has a relatively low oxygen permeability. Polystyrene and polyethylene are not suitable for steam-sterilizable probes because of low heat resistance.

For steam-sterilizable applications, changes in membrane properties with temperature must be reversible. Thin membranes (10–25 μm) have problems because they do not withstand large pressure differences, and the sterilization temperature (120 °C) causes an irreversible change in membrane thickness[26]. Borkowski and Johnson[21] used a 50 μm Teflon membrane, which could withstand about 20–30 times repeated steam sterilizations[141]. The steam-sterilizable probe of the Instrumentation Laboratory[26] employed a special double membrane to withstand sterilizations. The inner membrane was a 25 μm Teflon film and the outer membrane was 150 μm thick silicone which was reinforced by thin steel netting.

Figure 11 shows a plot of current output vs. the number of sterilizations for both types of probe. It is interesting to note that, while the current output of the polarographic probe (Instrumentation Laboratory probe) increased upon repeated sterilization, the current decreased for the galvanic probe (Borkowski-Johnson probe). Reduced membrane thickness is probably the reason for the current change of the polarographic probe, whereas consumption of the anode may be the main factor for the galvanic probe.

In bioreactor applications, *growth of microorganisms or soil coating* on the outer surface of the membrane is the most common cause for faulty measurements. The soil layer behaves as an additional membrane and the growth of microorganisms inhibit oxygen supply to the cathode. Frequent calibrations are required for contaminated probes. The contamination may be detected from the response of the probe since all the contaminated probes respond slowly[26].

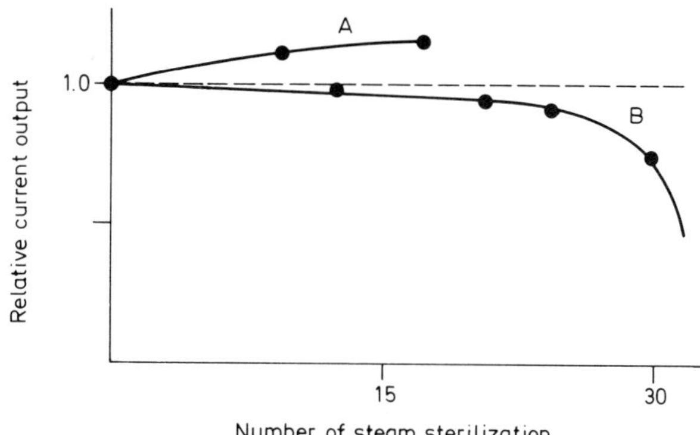

Fig. 11. Effect of steam-sterilization on probe current: polarographic probe (IL probe: Bühler and Ingold[26]) and galvanic probe (B–J probe: Tuffile and Pinho[141]). A: Instrumentation Laboratory probe B: Borkowski-Johnson probe

4.5 Instrumentation

Unless the current output is very large, some sort of current amplification is required for DO probes. Operational amplifiers which provide multi-range amplification and zero current suppression are normally used for this purpose. LaForce[85] described in detail the use of operational amplifiers for polarographic sensors. For an improved stability in current amplification, Orbisphere (York, Maine) employed chopper-stabilized amplifier in its DO meter[117]. Commercially available DO meters also incorporate temperature measurement and/or compensation circuits.

Often, existing current amplifiers, voltmeters or strip-chart recorders are directly connected to DO probes. For polarographic probes, a bias voltage of 0.6–0.8 V (depending on the type of electrode) has to be externally applied as shown in Fig. 5. The selection of a polarization voltage is discussed in Sect. 4.6. A mercury battery is preferred as the voltage source because the voltage discharge characteristics are superior to other batteries. For galvanic probes, usually a resistor is connected in series with the probe and the voltage drop across the resistor is monitored with a voltmeter or a potentiometric recorder.

For both the polarographic and the galvanic probes, the input impedance of the current measuring circuit has to be low so as not to affect the cell potential of the probe[142]. The value of series resistance is shown to affect the sensitivity and the response time of the galvanic probes[61, 79]. The galvanic electrode pair spontaneously generate a voltage of around 0.7 V which is sufficient for oxygen reduction at the cathode. When a resistor is connected in series with the probe (Fig. 12a), the oxygen current flows through the resistance and causes a voltage drop, V, across the resistor:

$$V = IR_L \quad , \tag{34}$$

where I is the oxygen current and R_L is the combined resistance of the parallel resistor and the input impedance of the measuring device. When the value of R_L becomes large, the increased voltage drop will significantly affect the cell potential. This is equivalent to change of polarization voltage in polarographic probes. Evidently, the probe calibration and the response will be affected. This effect will be more pronounced for large cathodes because the current output is high. The load resistance, R_L, has to be selected so that the voltage drop V is negligible compared with the cell potential. The same principle applies to polarographic probes.

A current meter (or amplifier) with zero input impedance is ideal for DO probes but a reasonably low value of R_L is acceptable. For example, if $R_L = 100\ \Omega$, the Mancy probe shown in Table 2 will give a voltage drop across R_L of 1 mV under air-saturated condition, which is negligible compared with 0.7 V cell potential. The probe would not function if the load resistance, R_L, is increased to 100 kΩ because the voltage drop due to R_L becomes 1 V, which is even higher than 0.7 V. Accord-

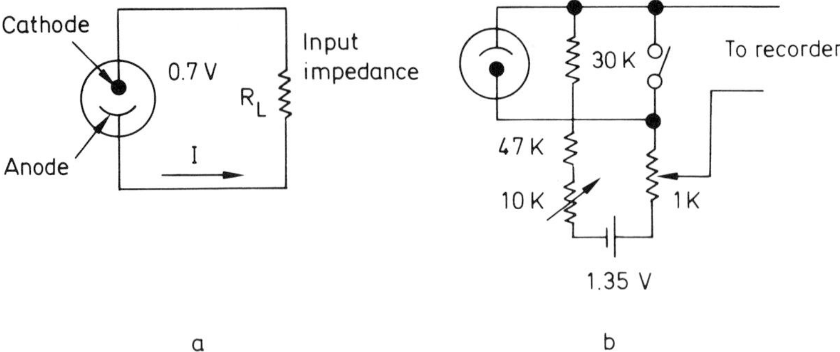

Fig. 12. (a) Equivalent circuit for oxygen electrode; (b) circuit used for measuring small changes in oxygen tension (Lipner et al.[97])

ing to Eq. (34), R_L may have high values when measuring low oxygen tension and when the cathode is small (since small cathodes produce low current).

Lipner et al.[97] showed that a simple bucking circuit shown in Fig. 12b could be used with the Mancy probe to measure very small changes in oxygen tension of a sample during enzyme reactions. Even if the residual current of the Mancy probe was relatively high, it could be effectively suppressed by an imposed opposing current.

4.6 General Design Considerations

One of the important considerations in DO probe design is to minimize the *residual current,* which is defined as the current output of the probe at zero oxygen level. Sometimes, it is called the background, dark, zero, offset or nitrogen current. Krebs and Haddad[83] listed four major sources which contribute to this residual current:

electrochemically active impurities in the electrolyte;
electrical leakage;
incorrect polarizing voltage;
and back diffusion of oxygen.

The effect of *reducible* or *oxidizable impurities* in the electrolyte is normally of short-term duration because they are scavanged by the cathode at the early stage of operation. Electrical leakage between the anode and the cathode through the insulating material is not normally significant for large cathode but becomes a problem for small cathodes. For example, the insulation requirement of the microelectrode shown in Table 2 has to be better than 7×10^{10} Ω to have a residual current of less than 1×10^{-11} A (resistance = polarizartion voltage/current). This is why glass is used as the insulating material for small cathodes. Epoxy resins, unless specially selected for their water resistance, often absorb water and increase current by forming an extraneous conducting path between the cathode and the anode.

The correct choice and the *stability of polarizing voltage* is important for stable performance of the probe. Figure 13 from Krebs and Haddad[83] shows three polarographic curves. If the polarizing voltage is chosen between points marked B and C and is main-

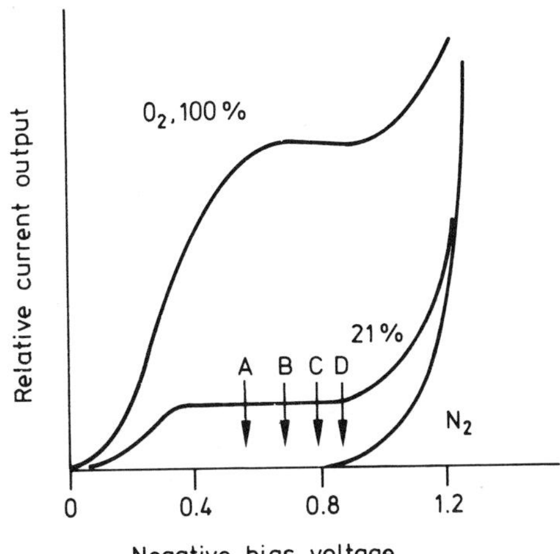

Fig. 13. Polarographic curves from a Pt-Ag/AgCl probe (Krebs and Haddad[83])

tained in this region, the probe current will depend only on the oxygen tension. However, when the polarizing voltage is larger than that indicated by point D, there will be residual current due to electrochemical reduction of water. If the polarizing voltage is less than that marked A, then the calibration of current vs oxygen tension will become non-linear. Since the shape of the polarographic curve changes depending on the electrode metal, the electrolyte concentration[29] and the size of the cathode[73], the optimum polarizing voltage has to be selected for each probe, separately. In general, large cathodes have wide and well-developed plateaus in their polarogram, whereas small cathodes have narrow and, often, very poorly-defined plateaus[73]. Probes with large cathodes, notably the galvanic probes designed to drive meters directly, may change their effective polarizing voltage if the combined resistance of the meter and the series resistor is relatively large. This was discussed in Sect. 4.5.

The fourth contributor to the residual current, namely, *back diffusion of oxygen* from the internal electrolyte is most pronounced for small cathodes. This is because the oxygen reduction due to back diffusion mainly occurs at the edge region of the cathode and, for small cathodes, the fraction of total area involved is relatively large. Mechanical designs which favor increased path length between the cathode and the electrolyte reservoir, as well as decreasing the thickness of the electrolyte film between the membrane and the cathode, minimize this effect[83]. Sometimes, oxygen dissolved in the plastic components of a probe diffuse back to the probe face when the probe is moved from an oxygen-containing environment to one free of oxygen[47]. Metal components for the sensor body reduce this type of back diffusion.

Mechanically well-designed probes with the above considerations in mind have very low residual current and are capable of measuring dissolved oxygen down to parts per billion (ppb) range. Orbisphere Model 2711 analyzer[117], which utilizes polarographic probe, has a 10 ppb full-scale range. Rexnord Model 3400 DO analyzer[123] employs a galvanic probe and has a 20 ppb full-scale range. Both probes are claimed to be extremely stable and rugged (for example, Orbisphere claims ±3% drift per year). These probe can be applied in studying the effect of low dissolved oxygen tension on the metabolism of facultative microorganisms[61].

Galvanic probes have been preferred in bioreactor applications[3, 59, 141, 142] but the disadvantage is the expendable nature of the anode. Currently available designs of the steam-sterilizable galvanic probe could be improved considerably by further decreasing the cathode area, increasing the anode area and increasing the path between the electrolyte reservoir and the cathode. A well-designed polarographic probe was shown to withstand repeated steam-sterilization with a high reliability[83].

It would be fairly simple to fabricate working DO probes in the laboratory provided that proper design considerations are taken into account. The advantage of making the probe in the laboratory is that the probe can be designed to meet the requirements of the specific application.

5 Operation of Electrodes

5.1 Calibration

DO probes can be calibrated in three ways: % saturation, partial pressure of oxygen and actual concentration. Since the reliability of measurements depends on calibration, it is vital that the calibration be done as accurate as possible. In all calibrations, good temperature control (±0.1 °C) of the test medium is required because the probe sensitivity changes markedly with temperature as will be shown in Sect. 5.3.

5.1.1 Calibration Based on % Saturation

Gas-phase calibration is rapid and convenient. The probe output current in nitrogen is set at 0% saturation while that in air is set at 100%. Oxygen can be used instead of air. This method is simple but recommended only for a rough calibration. Given the same oxygen partial pressure, the current output of the probe is lower in liquid phase compared with that in gas phase, because there exists a stagnant liquid film around the membrane even at high velocity. Equation (17) predicts such behavior.

For a better accuracy, the calibration has to be made directly in the liquid of interest. Air saturated liquid can be prepared by bubbling air through the solution at the desired temperature. A solution with zero oxygen content can be prepared either by stripping oxygen with nitrogen or by using chemicals such as sodium sulfite or sodium dithionite. These chemicals react rapidly with dissolved oxygen to provide anaerobic conditions. Addition of 2 volumes of saturated sodium sulfite to 100 volumes of liquid is sufficient for oxygen removal[37]. Dry powders can be directly added to a concentration of 0.01 M or higher for the same purpose[14].

5.1.2 Calibration Based on Partial Pressure

DO probes are known to measure the activity or the equivalent partial pressure of dissolved oxygen but not the actual concentration[106]. For example, the current output from the oxygen probe is the same for air-saturated 1 M KCl and air-saturated pure water, whereas the actual oxygen concentration in 1 M KCl is only 73% of that in pure water. In practice, the current output changes slightly depending on different liquids but this tendency becomes negligible when the cathode area decreases[131]. The calibration based on % saturations can be directly converted to partial pressures when the barometric pressure is known.

When water is equilibrated with air at temperature T, the partial pressure of oxygen, pO_2, is expressed as follows[53]:

$$pO_2 = [pB - p(H_2O)] \times 0.2095 \ , \tag{35}$$

where pB = temperature-corrected barometric pressure, $p(H_2O)$ = vapor pressure of water at a given temperature, 0.2095 = fraction of oxygen in atmospheric air.

The values of pB and $p(H_2O)$ can be obtained from Table 5. For example, when the barometric reading is 750 mmHg at 20 °C, the current output of the probe at 100% saturation corresponds to

$$pO_2 = [(750 - 2.44) - 17.515] \times 0.2095$$
$$= 152.9 \text{ mm Hg} \ .$$

Note that this corrected value is slightly lower than that of the uncorrected one (750 x 0.2095 = 157.1 mm Hg).

Table 5. Temperature corrections for mercury barometer and vapor pressure of water (Gelder and Neville[53]) (All units in mm Hg)

Temp. °C	740 mm	750 mm	760 mm	770 mm	$p(H_2O)$
18	2.17	2.20	2.23	2.26	15.460
19	2.29	2.32	2.35	2.38	16.460
20	2.41	2.44	2.47	2.51	17.515
21	2.53	2.56	2.60	2.63	18.631
22	2.65	2.69	2.72	2.76	19.807
23	2.77	2.81	2.84	2.88	21.047
24	2.89	2.93	2.97	3.01	22.356
25	3.01	3.05	3.09	3.13	23.734
26	3.13	3.17	3.21	3.26	25.185
27	3.25	3.29	3.34	3.38	26.715

5.1.3 Calibration Based on Concentration

The calibration based on partial pressures can be convered to concentrations if the solubility of oxygen in the liquid is known. The solubility is often expressed as the Bunsen coefficient, α, which is defined as ml O_2 absorbed by 1 ml of solvent at 0 °C and 1 atm pressure of O_2 [10]:

$$a = \frac{V_g}{V_s} \frac{273.15}{T} , \tag{36}$$

where V_g, V_s, and T are volume of gas absorbed, volume of the absorbing solvent and absolute temperature, respectively. For a spargingly soluble gas such as oxygen, pressures can be converted to concentrations by using Henry's Law:

$$c = p_o/H , \tag{37}$$

where H is the Henry's law constant. The conversion of a to H is as follows[10]:

$$H = \frac{22{,}414(760)}{1000a} . \tag{38}$$

The values of a for oxygen in water and other solvents are given in the literature[132]. The dissolved oxygen concentration changes markedly when the salt concentration in water increases. Such data are available in the literature[132] and a calculational method is also available[35]. However, a direct measurement of solubility is necessary when liquids of unknown composition are involved, or when the solubility data of the liquid are not available. The Winkler method[7] has been used widely for determining oxygen solubility in waters and wastewaters. Biochemists often use mitochondrial oxidation of NADH as the basis for DO probe calibration[14]. The oxidation of phenylhydrazine by ferricyanide was also used recently[109]. A DO probe method is given in Sect. 7.2.

In addition to these chemical methods, the mass spectrometric method, gas chromatographic method and manometric or volumetric method have also been used[10]. The manometric method of Van Slyke and Neill or that of Scholander[125] has been widely used for blood oxygen measurements and the advantage of this method for oxygen probe calibration is given by Elsworth[42]. Once the oxygen concentration in the solution is determined, then the probe output can be calibrated in terms of concentration. It has to be noted that the probe calibrated in one liquid cannot be used for measuring oxygen concentration in other liquids because oxygen solubility changes depending on the liquid composition.

5.1.4 Calibration for Long-Term Continuous Measurement

Probe calibration becomes a problem when it is necessary to monitor the dissolved oxygen continuously during a long-term cultivation. One problem is the drift of the probe sensitivity which occurs at approximately 2% per week[77]. Mclennan and Pirt[107] reported that after 5 weeks of continuous monitoring in cultivation media, the oxygen probe showed marked decrease in sensitivity and the calibration became non-linear at high dissolved oxygen tensions, although the residual current of the probe remained unchanged.

Recently, a DO probe capable of in situ calibration was introduced[81]. Kjaergaard[77] showed a simple method for recalibrating a DO probe during a long-term process by utilizing the known aeration capability of the bioreactor.

5.2 Response Time

The transient response of the DO probe becomes important when it is required to measure relatively fast concentration changes. Examples are: dynamic measurements of the volumetric oxygen transfer coefficient in bioreactors[8, 15, 40, 91]; oxygen respiration rate measurements in microbial suspensions[63]; concentration fluctuation measurements[87], low-level DO measurements[79], respiratory oxygen measurements[39, 51], etc. Because of these wide applications, the dynamic response of the DO probe has been studied extensively[15, 17, 57, 62, 80, 90–96, 124, 147].

The response time of the probe is the time required for the probe to reach a certain fraction of the steady-state current to a step change in oxygen concentration. The fractional response, Γ is experimentally determined as follows:

$$\Gamma = \frac{I_t - I_0}{I_s - I_0} , \tag{39}$$

where I_0 is the current output at time zero. Frequently, 90 or 95% response time is quoted in the literature.

Theoretically, the response characteristics of the probe can be estimated by using an electrode model. For a single layer model shown in Sect. 3.2, the fractional response of the probe to a step change in oxygen tension is given as follows[15, 62, 91]:

$$\Gamma = 1 + 2 \sum_{n=1}^{\infty} (-1)^n \exp(-n^2 kt) , \tag{40}$$

where k is the probe constant defined by Eq. (11). Linek rearranged Eq. (40) as follows:

$$\Gamma = 1 - 2 \exp(-kt) + 2 \sum_{n=2}^{\infty} (-1)^n \exp(-n^2 kt) \tag{41}$$

and showed that, for $\Gamma \geqslant 0.4$, the infinite series term can be neglected and k is the negative slope of a plot of $\ln(1-\Gamma)$ *vs.* t.

The factors affecting the response time of the probe can be estimated from Eqs. (11) and (40). A thinner membrane with a high oxygen diffusivity is expected to give a rapid response time. Evidently, varying d_m rather than D_m is more effective in obtaining a fast response. When more than one layer is involved in oxygen diffusion path, d_m in Eq. (11) has to be replaced with an effective membrane thickness $\bar{d}_t$ defined by Eq. (20). According to Eq. (20), a thick-liquid film with D_L comparable to D_m, is expected to slow down the probe response. In other words, the response time of the probe increases with decrease in liquid velocity (Fig. 14a) and with increase in liquid viscosity. The latter holds from the following relationship for Newtonian fluids[150].

$$\frac{D_1 \mu_1}{T_1} = \frac{D_2 \mu_2}{T_2} , \tag{42}$$

where μ is viscosity and D is diffusivity.

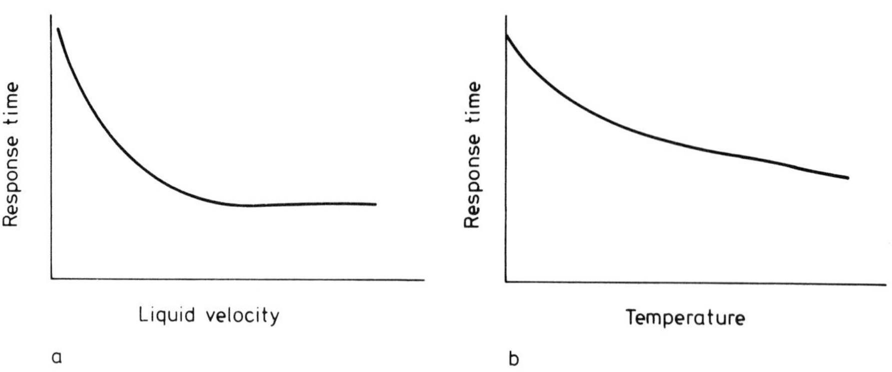

Fig. 14. Probe response time as a function of (a) liquid velocity and (b) temperature (Kimmich and Kreuzer[75)])

The temperature also affects the response time since the oxygen diffusivity of the membrane, D_m, is a function of temperature[16)]:

$$D_m = D_m^* \exp(-E_D/RT) \ , \tag{43}$$

where D_m^*, E_D, R, and T are diffusivity at base temperature, activation energy for diffusion, gas constant and absolute temperature, respectively. Equation (43) shows that the diffusivity increases with increase in temperature. Schuler and Kreuzer[128)] showed that 100 μm diameter cathode with a 12 μm Teflon membrane gave 95% response time of 2 s at 20 °C, but the response time decreased to 1.3 s at 50 °C (Fig. 14b). Severinghaus[130] operated a probe at an elevated temperature to obtain a fast response.

Often, DO probes show slowdown in the last 20% of the response to a step change in oxygen tension[15, 80, 93)] and a marked difference in response time between the upstep response and the downstep response (Fig. 15). This phenomenon was observed for the Mackereth probe[61)], Yellow Springs Instrument probe[15)], Borkowski-Johnson probe[93)], and other probes[80, 123)]. These probes commonly employed a large pool of electrolyte very close to the cathode. The tailing of response was explained in terms of lateral diffusion of oxygen from the electrolyte reservoir[80)]. Probes of other de-

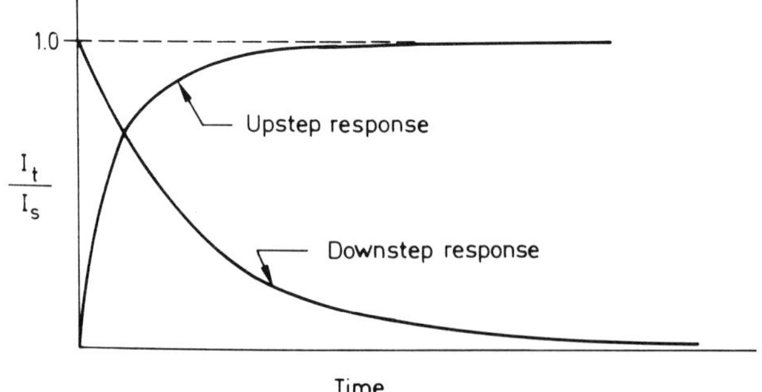

Fig. 15. Response hysteresis of oxygen electrodes (Harrison[61)])

signs[93, 106] did not show such behavior. These probes either had small amounts of electrolyte or the electrolyte reservoir was located far from the cathode.

Several authors presented electrode models which take into account of the lateral diffusion[69, 80]. Linek and Benes[93] used a multi-layer, multi-region model for describing the tailing response but stated that such probes are not convenient because more than three parameters are needed for describing the dynamic behavior of the probe.

With a 25 μm Teflon or polypropylene membrane, 95% response time of 15 s is usually obtained in water at room temperature (Table 2). Depending on the probe design, the response time can be further decreased without affecting the condition of membrane-controlled diffusion: for example, the ring-shaped cathode of Kimmich and Kreuzer[75] had a 95% response time of only 0.4 s. Microprobes have very fast response time (0.1 s for 95% response) due to a very thin membrane. Davies[37] showed that 100% response time of the DO probe can be as fast as 0.5 ms. Slow response can be made faster or compensated for electronically[22, 84].

When using a DO probe for measuring unsteady concentration change, it is important to check the probe response before and after use to prevent measurement errors. The response check has to be made in the same liquid as the test medium under similar hydrodynamic conditions. This is also a good way of checking bad or aged probes.

5.3 Effect of Temperature

A number of authors[21, 104, 106, 131] reported 1-5% increase in probe sensitivity per °C for various DO probes. Evidently, the measurement error would be large if the temperature of the medium is not well controlled. In addition to the temperature coefficient of the probe, the oxygen solubility changes with temperature. For example, in water, the oxygen solubility changes approximately 2%/°C at 25 °C. If the temperature coefficient of the probe is assumed 3%/°C, the combined error in calculating oxygen concentration will be greater than 5%/°C in the worst case. Therefore, good temperature control of the medium (±0.1 °C or better) is essential for accurate measurements.

The effect of temperature on probe sensitivity can be estimated by using the Arrhenius relationship for the diffusivity, D_m [Eq. (43)] and the solubility, S_m, of the membrane[16]:

$$S_m = S_m^* \exp(-\Delta H/RT) \; , \tag{44}$$

where ΔH is the heat of solution. Since the permeability, P_m is a product of solubility and diffusivity, Eqs. (43) and (44) are combined as follows:

$$P_m = P_m^* \exp(-E/RT) \; , \tag{45}$$

where

$$E = \Delta H + E_D \; .$$

The steady-state current output of the probe given by Eq. (10) can be rewritten as follows:

$$I_T = NFA \frac{p_0}{d_m} P_m^* \exp(-E/RT) \; . \tag{46}$$

The activation energy, E, was 8.8 kcal $(\text{g-mole})^{-1}$ for polyethylene[16] and 7.8–9.6 kcal $(\text{g-mole})^{-1}$ for polypropylene membranes[91].

Under a given oxygen partial pressure, Eq. (46) can be rearranged as follows:

$$I_T = A_1 \exp(-A_2/T) \ , \tag{47}$$

where

$$A_1 = NFA(P_m^*/d_m)p_0 \ ,$$

and

$$A_2 = E/R \ .$$

Thus, when the logarithm of the current output is plotted against T^{-1}, a straight line with a slope of $-A_2$ is obtained. According to Eq. (47), the sensitivity increases with increase in temperature but the trend is not linear but exponential. Therefore, the practice of giving temperature coefficients of sensitivity per °C is basically incorrect, although it may be applicable for a narrow temperature range[16]. In Eq. (47), A_1 is a function of cathode area, membrane thickness and membrane permeability whereas A_2 depends only on the property of the membrane.

Temperature compensation of the probe sensitivity can be accomplished by several methods. The simplest and most accurate method is the use of calibration curves. The dissolved oxygen level can be directly read from the calibration curve corresponding to the temperature of the medium. This method is somewhat cumbersome and may not be applicable for continuous measurements.

Another method is an automatic compensation using a thermistor. The resistance of a thermistor R_T is given by the following equation[23]:

$$R_T = B_1 \exp(B_2/T) \ , \tag{48}$$

where B_1 and B_2 are constants for a given thermistor. If a thermistor is used as the cell load resistance and the voltage developed across it is measured using a voltmeter or a recorder with a very high input impedence (Fig. 16a), then the voltage measured, V_T, is:

$$V_T = I_T R_T \ . \tag{49}$$

From Eqs. (47)–(49), the voltage V_T, which is a measure of oxygen tension becomes as follows:

$$V_T = A_1 B_1 \exp(B_2 - A_2)/T \ . \tag{50}$$

If $B_2 = A_2$, then the measured voltage is independent of T. Thus, the temperature dependency of the probe sensitivity can be completely compensated.

In practice, the selection of a thermistor poses some problem because R_T is usually high: on the order of 5–500 K. A high value of R_T not only requires a recorder with very high input impedence but also adversely affects probe performance as described in Sect. 4.5. Morisi and Gualandi[112] circumvented this problem by using a compensating voltage source in the circuit (Fig. 16b), which resulted in a temperature compensa-

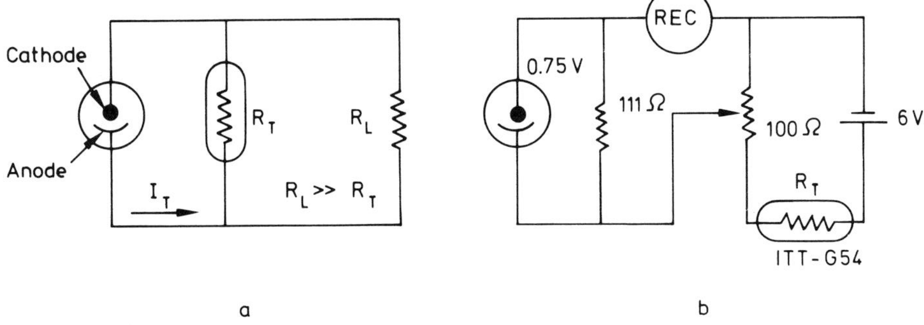

Fig. 16. (a) Principle of temperature compensation, (b) temperature compensation circuit (Morrisi and Gualandi[112])

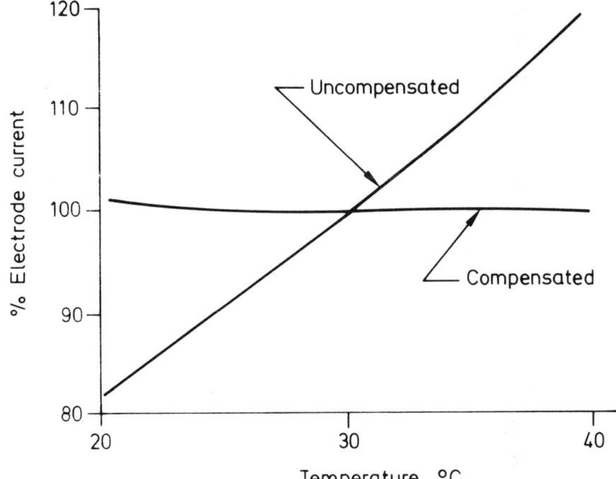

Fig. 17. Result of temperature compensation (Berkenbosch[17])

tion of 0.1%/°C in the range of 17–26 °C. Others[16, 23] gave temperature compensation circuits for polarographic probes. Figure 17 shows a result of temperature compensation.

It has to be noted, however, that the probe response characteristics change as a result of temperature compensation. An over-compensation occurs during the initial stage of transient response because the thermistor responds much faster than the probe[112].

A method commonly employed for temperature compensation in commercial units involves changing the degree of current amplification by using a thermistor in the feed back circuit of the amplifier[142]. Often, a separate knob is available for manual temperature compensation. Most of the commercially available probes incorporate a thermistor in the probe body for temperature measurement and compensation.

5.4 Effect of Liquid Film

The DO probe, when placed in a stagnant liquid, produces a diffusion gradient extending outside the membrane and farther into the liquid. The size of the steady-state diffusion field is proportional to the size of the cathode. When the liquid is stirred, the diffusion gradient can no longer be extended beyond the liquid film around the membrane. Since the diffusion gradient becomes steeper with decreasing liquid film thickness, the current output of the probe increases with liquid velocity (Fig. 18b). This so-called "flow dependency" of the probe sensitivity is higher for a probe with a larger cathode diameter (Fig. 18a) because the size of the stagnant diffusion field is proportionally larger than that of a small cathode.

For proper operation of the probe, the liquid has to be stirred beyond a certain level (v_c in Fig. 18a) in order to maintain membrane-control of oxygen diffusion. The critical velocity, v_c, of the liquid is the velocity where the probe output reaches 95% (95–99% depending on the definition) of the steady value. Of course, v_c depends on liquid viscosity but, given a liquid, v_c is smaller for smaller cathodes. For example, with a 25 μm Teflon membrane, a cathode of 5 mm diameter required v_c of 70 cm s^{-1} in water[3)], whereas only 5 cm s^{-1} was required for 25 μm diameter cathodes[45)]. Nomograms for assessing the critical velocity corresponding to different size cathode are given by several authors[3, 128)]. When the cathode diameter is less than 1 μm, the probe becomes insensitive to liquid flow even without the membrane[136)]. In this case, the diffusion field of the cathode is so small that it is always contained inside the minimum liquid boundary layer around the cathode. This is why such probes are capable of measuring local concentration rapidly with a high spacial resolution.

Theoretically, the effect of liquid film on probe performance can be estimated from Eq. (17). Given the same oxygen partial pressure, the current output in the gas phase, I_G and that in the liquid

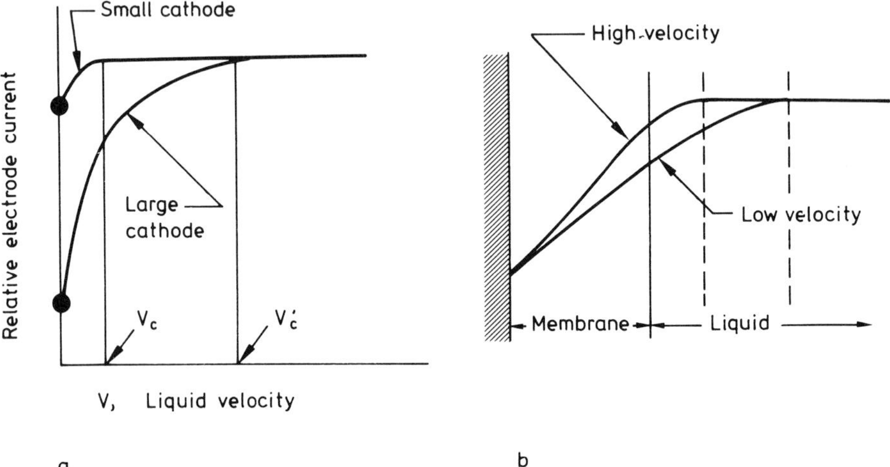

Fig. 18. Flow sensitivity of DO electrodes (a) effect of cathode size; (b) effect of liquid velocity

phase, I_L can be expressed as follows:

$$I_G = NFAP_m\left(\frac{1}{d'}\right) p_o \ , \tag{51}$$

$$I_L = NFAP_m\left(\frac{1}{d' + \frac{P_m}{P_L} d_L}\right) p_o \ , \tag{52}$$

where

$$d' = d_m + \frac{P_m}{P_e} d_e \ . \tag{53}$$

Thus,

$$\frac{I_L}{I_G} = \frac{1}{1+L} \ , \tag{54}$$

where,

$$L = \frac{P_m/d'}{P_L/d_L} \ . \tag{55}$$

The significance of the liquid film resistance can be determined by a single parameter L defined by Eq. (55), which is the ratio of mass transfer coefficient of the liquid film and that of the membrane (including electrolyte layer). L can be determined experimentally from the measured values of I_G and I_L:

$$L = (I_G/I_L)_{measured} - 1 \ . \tag{56}$$

Equations (54) and (55) show that the liquid film resistance becomes significant in viscous liquids since the oxygen permeability generally decreases with increasing viscosity.

Often there are situations where sufficient velocity of the medium cannot be provided under actual measurement conditions. Examples are *in vivo* blood oxygen measurements and DO measurements in deep waters such as lakes and rivers. When measuring the oxygen tension inside the microbial slime layers[28)] and in tissues[136)], the medium cannot be stirred at all. Microprobes have been used for the latter applications. Probes for use in deep water bodies employ small cathodes and thick membranes, and sometimes are equipped with stirrers.

Linek and Vacek[92)] presented an electrode model incorporating the liquid film resistance. Dang et al.[36)] introduced a simplified method of using the probe in very viscous media. This method is given in Sect. 7.1.

5.5 Handling, Maintenance, and Other Practical Considerations

Proper care and precaution are required in order to use a probe successfully. In general, the manual from the probe manufacturer provides adequate information. Some of the important points are given here.

When *refilling* the electrolyte solution, care has to be taken not to include air bubbles inside the electrolyte chamber, since the trapped bubbles interfere with the measurement. The membrane has to be tightly and securely fit over the cathode so that the electrolyte layer between the membrane and the cathode is minimized and kept constant. As shown by the electrode model in Sect. 3, the membrane and the electrolyte layer are directly related to the probe performance.

The *aging of the probe* can be detected by checking the probe sensitivity frequently. An excessive increase or decrease in current at a given oxygen tension indicates aging. A more convenient way of checking is by the transient response of the probe. Aged probes almost always show slower response. The loss of linearity in calibration and unstable performance are symptoms of aged probes. Aged probes can be rejuvenated by refilling the electrolyte and by cleaning the anode and cathode. The cleaning procedures are given in Sect. 4.2.

Special care has to be exercised in *handling steam sterilizable probes.* The most serious problem is the loss of electrolyte during the cooling cycle of autoclaving. If part of the electrolyte is lost so that the lead anode protrudes above the surface, then, after a few weeks, the anode breaks at the electrolyte surface. It is essential to maintain a slow cooling rate or to pressurize the autoclave above the atmospheric pressure during the cooling cycle in order to prevent the loss of electrolyte by evaporation or boiling[42)]. The probe has to be "conditioned" before starting measurements. The purpose of conditioning a probe is to de-oxygenate the electrolyte so that the current generated in the external circuit is a measure of oxygen entering through the membrane. Autoclaving usually conditions the probe but the anode and the cathode have to be shorted during and after steam sterilization to help remove and maintain oxygen-free condition in the electrolyte. Although a probe can normally withstand 15–30 times of autoclaving[26, 42, 141)], it was recommended to change the electrolyte after 5 or 6 sterilizations[120)]. The above discussions mainly apply to galvanic probes but similar precautions are necessary for polarographic probes.

The position of a DO probe in an air-sparged stirred tank or a bubble column affects measurement accuracy. Evidently, the probe has to be located at a position where the local liquid velocity is above v_c defined in Sect. 5.4. However, locating the probe close to the agitator creates the problem of direct contact between the bubbles and the probe membrane. The readings may be distorted unless the oxygen concentration in bubbles is in equilibrium with liquid. Linek and Vacek[94)] have shown that the probe reading varies depending on probe location in the tank. When the probe is used in a large bioreactor, placing the probe in stagnant water regions has to be avoided, not only because of the large liquid film effect, but because in these ranges microbial films are formed on the membrane surface, especially in highly viscous media. Linek and Vacek[94)] recommended locating the probe in a bypass loop connected to the stirred tank, to avoid bubbles contacting with the membrane. The probe may be inserted from the bottom[91)] or side of the vessel so that the probe membrane faces upwards. Sometimes, putting a suitable shield around the tip of the probe effectively prevents direct contact of bubbles with the membrane. Although the optimum location of the probe has to be determined case by case, the above guidelines have to be taken into consideration.

6 Sources of Error in Measurements

6.1 Errors due to Probe Characteristics

Improper calibration of probe may cause measurement errors. For polarographic probes, LeFevre et al.[89] showed that even a newly cleaned electrode (with 600-grade emery paper and ammonium hydroxide) showed a slight change in calibration after several hours. When the electrode is used for a long time without cleaning, the calibration changes rapidly with time and often becomes non-linear and unpredictable: sometimes the sensitivity increases[88] but sometimes it decreases[89]. Although the galvanic probe is generally more stable than the polarographic type, the calibration changes eventually[61, 107]. Calibrating the probe before and after use is recommended.

The calibration has to be made in the liquid where the actual measurements are to be made. Severinghaus[131] showed that the probe sensitivity decreased as much as 16% in 50% glycerine solution compared with that in water. Even small cathodes with thick membranes show changes in calibration in different media. For example, a Clark-type sensor with a 20 μm diameter cathode and a 20 μm polypropylene membrane, exhibited a 2–6% change in calibration between gas and blood[1]. This trend will be more significant with larger cathodes. When the liquid changes its property during measurement such as in batch cultivation, the calibration change has to be taken into account, especially when there is a marked difference in liquid viscosity between the initial and the final medium.

The temperature of the medium has to be controlled within ±0.1 °C both for the calibration and the actual measurement in order to achieve maximum accuracy. In calculating the actual dissolved oxygen concentration, two types of errors are involved: one is the temperature dependency of the probe sensitivity and the other is the change in oxygen solubility with temperature. With a poor temperature control, for example, ±0.5 °C, the measurement error could be as high as 10%[130]. Probes with automatic temperature compensation would be better for steady-state measurements but the transient behavior of the probe may not be good if the time response of the probe and that of the compensating circuit are not well matched. As mentioned earlier, over-compensation occurs for short times when a thermistor is used as a compensator[112]. When a different thickness membrane is used, the temperature coefficient of the probe has to be reestablished. The temperature compensation set for one membrane thickness may not hold for membranes of different thickness[91].

Liquid hydrodynamics is also a source of error in DO measurements. A thick liquid film around the membrane decreases the probe sensitivity and increases the response time of the probe. Linek and Vacek[92] showed that substantial error could be introduced in measuring the probe constant if the liquid film resistance is neglected. Both the calibration and the actual measurement have to be made under similar hydrodynamic conditions to avoid possible measurement errors. Whenever possible, the stirring speed or the liquid velocity has to exceed the minimum value beyond which the probe sensitivity is not affected. Methods of obtaining the probe constant in the presence of liquid film were given by Linek and Vacek[92] and Dang et al.[36].

As shown in Sect. 4.6, the correct selection and the stability of the polarization voltage are essential for good performance of the probe. The polarization or bias voltage has to be maintained constant. This is especially true when the polarogram has a narrow plateau. When the plateau is not well defined, which is the case for very small cathodes, a slight change in polarization voltage causes a considerable change in calibration. Carey and Teal[29)] showed that the shape of the polarogram changed depending on the electrolyte concentration, membrane material and membrane thickness. The loss of water from the electrolyte, the aging of the probe and the change in membrane property due to repeated autoclaving are likely to result in polarization voltage change and thus the instability of the probe. Sometimes, the polarization voltage is changed due to the parallel resistor used for measurement, especially when a large cathode is involved. As outlined in Sect. 4.5, the resistor or the input impedence of the amplifier has to be selected not to affect polarization voltage.

6.2 Errors due to Measurement Medium

When the probe is used in *polluted environments,* the cathode surface can be poisoned and this results in the deterioration of probe performance. Sulfurous gases such as H_2S, SO_2, and thio-organic materials are reported to be poisonous to cathodes[64)]: Ag is most sensitive and less so with Pt and Au. Pretreating the sample solution before using the probe is one solution. Another approach is to remove H_2S by a special membrane. If a lense tissue soaked in cadmium nitrite is sandwiched between the membrane and the cathode, H_2S can be effectively removed[64)]. Other gases that interfere with dissolved oxygen measurement include Cl_2, Br_2, I_2, and oxides of nitrogen.

When oxygen is consumed by reaction in the liquid film around the DO probe, measurement errors occur depending on reaction kinetics, bulk oxygen concentration and the thickness of the liquid film. Lundsgaard et al.[103)] studied this reaction effect by employing a steady-state mass balance inside the liquid film. They showed that, for a zeroth order reaction, the oxygen tension at the membrane surface, p'_m is given as follows:

$$p'_m = \frac{p_0 - \frac{1}{2}(k_0/P_L)\, d_L^2}{1 + (P_m/P_L)(d_L/d')} , \qquad (57)$$

where k_0 is the zeroth-order rate constant. When no reaction occurs in the liquid film, p_m becomes:

$$p_m = \frac{p_0}{1 + (P_M/P_L)(d_L/d')} . \qquad (58)$$

Equation (57) shows that if there is a reaction inside the liquid film, the partial pressure at the membrane surface will become lower than that without reaction [Eq. (58)]. The relative error can be found by dividing Eq. (57) by Eq. (58):

$$\frac{p'_m}{p_m} = 1 - \frac{1}{2}\frac{d_L^2\, k_0}{P_L\, p_0} \quad . \tag{59}$$

For fixed values of liquid film thickness, d_L, and liquid oxygen permeability, P_L, the percentage error will become high at low p_0 and at high k_0. Given $k_0 = 20\ \mu M\ s^{-1}$, $d_L = 30\ \mu m$[103], and $P_L = 7.075\,(10^{-7})\ ccO_2\ cm^{-1} s^{-1}$ at 760 mm Hg[128], the oxygen partial pressure of the bulk liquid has to exceed 6 mm Hg in order to have an error less than 5%. This corresponds to approximately 4% saturation of oxygen at room temperature. Lundsgaard et al.[103] calculated errors for Michaelis-Menten kinetics and concluded that the oxygen probe can not work in dense microbial cultures with K_M (Michaelis-Menten constant) lower than 1 μM, when bulk concentration is in the order of K_M.

When an oxygen microelectrode is used for measuring the oxygen concentration gradient inside the *microbial slime layers*[28] and the stagnant liquid layers[86], the following questions arise: "How local is this measurement?" and "Does the oxygen consumption by the probe disturb the existing concentration gradient?". According to Grünewald[58], the microprobe measures local concentration without affecting the existing concentration gradient outside the membrane only when the diffusion gradient due to cathode reaction is entirely confined inside the membrane (Fig. 19a).

When the diffusion gradient of the cathode extends outside the membrane, the existing concentration gradient in the medium changes. In this case, the probe measures the concentration not at the surface of the membrane but at some distance away from

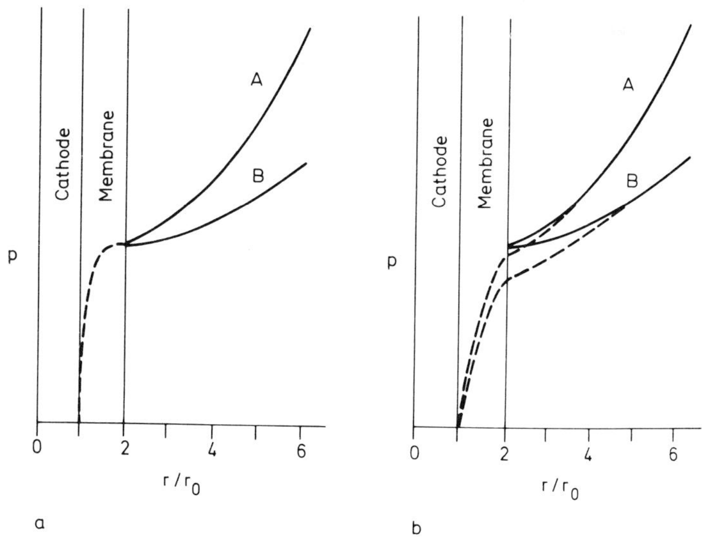

Fig. 19. Spacial resolution of microelectrode: (a) probe with membrane-controlled diffusion (b) measurement error due to loss of membrane-controlled diffusion (Grünewald[58])

the membrane. As shown in Fig. 19b, the measurement error becomes smaller when there is a steeper gradient (A in Fig. 19b) outside the membrane. A steeper gradient means a large flux of mass and the fraction consumed by the probe will become smaller compared with that of a less steep gradient (B in Fig. 19b). Therefore, the maximum error will occur when there is no oxygen concentration gradient outside the membrane. This case was dealt with in Sect. 3.3. The condition for an accurate measurement of local concentration was given by Eq. (31):

$$\frac{P_0\ d_m}{P_m\ r_0} \geqslant 99\ .$$

When $P_0 = P_m$ (this is the worst case), or in media with very low oxygen permeability, the cathode diameter has to be less than 0.02 μm in order to measure oxygen tension at 1 μm in front of the cathode. The situation becomes better when P_0 is considerably larger than P_m. When a polystyrene-covered microcathode is used in water at 25 °C, the cathode diameter can be increased to 0.4 μm in order to have the same spacial resolution[86]. Silver[135] showed experimentally that the spacial resolution of microelectrodes was better than those predicted by mathematical models[6, 58]. An excellent guideline for selection and calibration of microelectrodes was given by Silver[135].

7 Applications

7.1 Measurement of k_La and Respiration Rate

Since the introduction of dynamic measurement technique by Bandyopadhay and Humphrey[8], DO probes have been widely applied in measuring the aeration capacity of bioreactors and wastewater treatment units[15, 36, 62, 63, 90–96, 115, 124, 147]. The aeration capacity of a given vessel is characterized by the *volumetric mass transfer coefficient*, k_La, which is the most important parameter in scale-up of aeration devices. Although other methods[40] have been used in measuring k_La, the DO probe method is simpler and more convenient. However, some precautions are required in applying this method as will be discussed below.

The method involves measuring the DO concentration in the liquid phase to a step change in gas concentration. The dissolved oxygen is first removed, usually by sparging with nitrogen, and the change in DO is monitored with the probe after the resumption of oxygen supply. In this case, the oxygen mass balance in the liquid phase gives the following expression:

$$\frac{dc}{dt} = k_La(c^* - c)\ , \tag{60}$$

where c^* is the saturation value of DO at the gas-liquid interface and c is the concentration in the liquid bulk. If the DO probe can follow the concentration change instantaneously, then k_La is simply the negative slope of a plot, $\ln(c^* - c)$ *vs* t. But, since the probe has a finite response time, the above method leads to large errors.

Recently, a number of authors[15, 62, 90, 147] used electrode models for evaluating k_La. The mathematical treatment was essentially the same as that for obtaining the step response of the probe [Eq. (40)] except that an exponential rather than a step change in concentration was used as the boundary condition. The normalized probe response to the resumption of aeration was derived as follows[90]:

$$\Gamma = 1 - \frac{\pi\sqrt{B}}{\sin(\pi\sqrt{B})} \exp(-k_Lat) - 2 \sum_{n=1}^{\infty} (-1)^n \frac{\exp(-n^2 kt)}{n^2/B - 1} , \tag{61}$$

where

$$B = (k_La)/k ,$$

and k is the probe constant defined by Eq. (11).

The value of k_La can be obtained from the experimental data by the nonlinear least squares fitting method[62, 147]. Linek[90] showed a simplified method of obtaining k_La: the infinite series terms in Eq. (61) becomes negligible beyond a certain value of kt, and k_La can be determined from the slope of a plot, $\ln(1 - \Gamma)$ vs time. In estimating k_La, Wernau and Wilke[147] used a generalized graph prepared from the probe response equation. Their method required only the slope at the inflection point of a plot, Γ vs t.

The above methods are applicable when the liquid film resistance around the probe membrane is negligible. However, the methods lead to errors when there is a large film resistance. Such is the case when the liquid velocity around the probe is insufficient or when the liquid is viscous. Linek and Vacek[92] incorporated the liquid film resistance in the electrode model and presented a probe response equation for calculating k_La. The parameter L defined by Eq. (55) was used to represent the liquid film effect. Linek and Benes[93] showed that probes exhibiting tailing response were inconvenient for k_La measurement since they require additional parameters for describing the probe dynamics. The mathematical expression for the probe response became progressively complicated with increase in the number of parameters.

The methods of evaluating k_La based on electrode models are somewhat complicated mathematically and often require numerical calculations by the computer. A simpler method was introduced recently[36, 115] which enabled estimating k_La directly from the probe response curves. Nikolaev et al.[115] showed that the area between the normalized step response curve and the normalized aeration response curve (Fig. 20) is equal to $1/k_La$:

$$\int_0^\infty (1-\Gamma)_{aeration}\, dt - \int_0^\infty (1-\Gamma)_{step}\, dt = \frac{1}{k_L a} . \tag{62}$$

They reported a good agreement between this method and the Winkler method. This type of parameter estimation is called the "moment method". Linek and Benes[96] showed that this method gave correct estimation of k_La even with probes showing tailing response. The other advantage is that the method can be used in viscous media because the liquid film effect is cancelled out in estimating k_La when both the step response and the aeration response are obtained under the same hydrodynamic conditions.

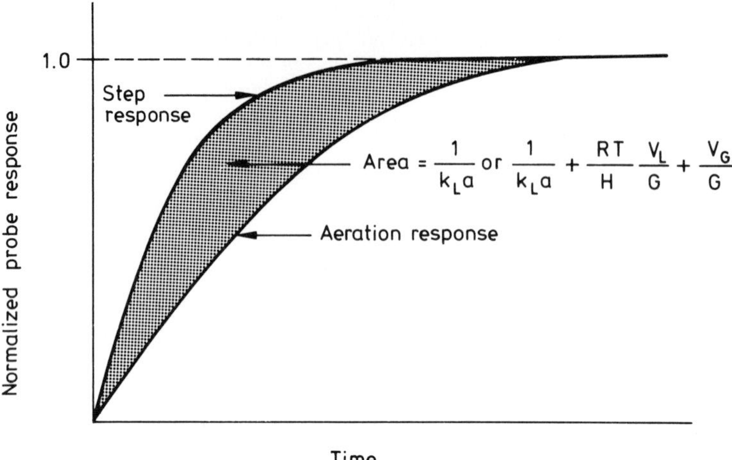

Fig. 20. Graphical method of determining k_La (Dang et al.[36], Nikolaev et al.[115])

Another possible source of error in estimating k_La, *the gas dynamics,* was recently identified by Dunn and Einsele[40]. When the dissolved oxygen is removed by sparging with nitrogen, the dispersed gas phase is pure nitrogen before the resumption of oxygen supply. During the initial stage of aeration, the nitrogen dilutes the gaseous oxygen and is displaced at a rate depending on the mean gas residence time. The measurement error due to this gas dynamics is especially significant in the case of intensely agitated tank reactors, which exhibit a well-mixed gas phase. In order to take the gas dynamics into account, Dang et al.[36] employed an oxygen balance in the gas phase assuming well-mixed dispersed gas phase:

$$\frac{dc_g}{dt} = \frac{c_g' - c_g}{V_G/G} - k_La\,(c^* - c)\frac{V_L}{V_G}\,, \tag{63}$$

where c_g, c_g', V_G, V_L, and G are oxygen concentration in the dispersed gas phase, inlet gaseous oxygen concentration, the volume of the dispersed gas phase, liquid volume and the gas flow rate, respectively. By solving Eq. (63) simultaneously with Eq. (60) and two equations describing probe dynamics, they derived the following relationship:

$$\int_0^\infty (1-\Gamma)_{aeration}\,dt - \int_0^\infty (1-\Gamma)_{step}\,dt = \frac{1}{k_La} + \frac{RT}{H}\frac{V_L}{G} + \frac{V_G}{G}\,, \tag{64}$$

where R and T are gas law constant and absolute temperature, respectively. As shown in Fig. 20, the shaded area corresponds to $1/k_La$ plus the correction terms given in Eq. (64). The method was successfully applied in evaluating k_La in very viscous media.

Linek and Vacek[94] also showed that a substantial error could be introduced in the estimation of k_La when the effect of start-up period (the period of time that will elapse before the new concentration level is established and/or before the k_La regains its steady-state value after the resumption of oxygen supply) was neglected. Expressions for the probe response were derived by assuming linear change of the pertinent characteristics during the start-up period.

Votruba et al.[143, 144] studied the *effect of air bubbles* on k_La determination by the dynamic method. They stated that when the bubbles hit the probe membrane frequently, the probe measures not the bulk concentration but a concentration between the equilibrium value and the bulk value. They used a concept of local gas hold-up in estimating the actual concentration from the probe reading, and presented an equation for calculating k_La. However, since the standard DO probes are hit mainly by large bubbles, the estimation of local gas hold-up by the probe may not be adequate and thus the use of their method requires caution. There have been some arguments on the application of this method[95, 145].

The moment method of obtaining k_La (Fig. 20) is in general more convenient and accurate compared with the methods based on electrode models. An added advantage is that the method can be applied even in very viscous media such as in dense cultures. However, in order to apply this method successfully, the experimental conditions such as aeration rate, stirring rate, liquid phase composition and the electrode location, have to be the same for both the step response and the aeration response experiments[36]. Also, the probe location has to be selected to avoid direct contact of the probe membrane with the bubbles. In obtaining the step response, Dang et al.[36] recommended moving the probe rapidly from a nitrogen stream into oxygen saturated liquid.

The *respiration rate,* R, in microbial suspensions has been measured with DO probes. The method involves monitoring the probe response after a sudden interruption of oxygen supply[8]. The rate of change in dissolved oxygen is then equal to the respiration rate:

$$\frac{dc}{dt} = -R \,. \tag{65}$$

When applying this method, it is important not to let the dissolved oxygen concentration go down beyond the critical value, below which the respiration of microorganisms is damaged[63, 91]. Also, the oxygen supply through the free surface has to be negligible.

7.2 Other Applications

Aiba et al.[2, 4] and Berkenbosch[17] used DO probes for measuring oxygen permeability and diffusivity through *polymer membranes.* The method involved measuring the probe covered with a test membrane to a known oxygen partial pressure. P_m was measured from Eq. (10) with known values of N, F, A, d_m, and p_0. Equation (8) was used for calculating D_m via three different methods: the direct estimation by Eq. (8), the moment method and the slope method. The moment method involved using the area under the normalized response curve, which is similar to that used for k_La measurement. For these applications, the cathode diameter of the probe has to be large compared with the thickness of the membrane because Eq. (10) was derived under the assumption of one-dimensional diffusion.

Oxygen diffusivities of liquids have been estimated by monitoring the response of the microprobe after a step change in oxygen partial pressure in the gas phase[46, 55, 86]. In this case, the depth of liquid has to be small to prevent convenction. The analysis is similar to that given by Eq. (1) and its boundary conditions. This method was used for estimating the oxygen diffusivity in *bovine serum albumin solutions* and in electrolyte solutions[55].

The oxygen microprobes have been used in measuring the local concentration profiles in *microbial slime layers,* from which the apparent oxygen diffusivity was calculated[28]. The surface renewal rate in an agitated liquid was measured with the microprobe by Bungay et al.[27] and Tsao et al.[139]. The oxygen microprobe was also shown to be able to measure the length and the velocity scales of liquid movement at the liquid surface region[86]. This method was applied in showing the changes in hydrodynamic parameters at the gas-liquid interfacial zone with the added surface active materials such as surfactants and proteins.

DO probes measure oxygen partial pressure in the liquid. Therefore, solubility data are needed for conversion to concentrations. Several methods are available for measuring oxygen solubility. One method involves stripping of dissolved oxygen in an air-saturated liquid by sparging with inert gas (H_2 or CO_2) and monitoring gas phase oxygen[47]. The procedure can be used for determining oxygen solubility in as little as 0.1 ml water. Fatt[47] used a simpler approach by monitoring oxygen tension change after adding a small sample of test liquid in a closed flask which was previously filled with water of known oxygen solubility. For small changes in oxygen tension in the flask, the Henry's Law coefficient for the contents of the flask can be considered to remain constant. Therefore, the change in oxygen tension, p, can be taken to be proportional to concentration change:

$$S_L = S_w \frac{p_2 - p_1}{p_1 - p_0}, \qquad (66)$$

where S_w = oxygen solubility in water, p_0 = initial oxygen tension in flask, p_1 = oxygen tension after adding 0.25 ml of air-saturated water, p_2 = oxygen tension after adding 0.25 ml of air-saturated sample.

Since the ratio of sample volume to flask volume used was 1×10^{-3}, the solubility of the water in the flask could be assumed to remain constant even after adding the sample. The accuracy of this method was comparable with other methods.

Often, data on *oxygen solubilities of cultivation media* is needed to convert partial pressures into actual concentrations by Henry's Law. Although a correlation is available for estimating oxygen solubility in an electrolyte solution[35], the method gives an estimation errors of −8% to +12%[98]. Besides, the culture medium contains non-electrolytes such as glucose, alcohols, hydrocarbons, etc. Liu et al.[98] used a DO probe for measuring oxygen solubility in actual medium by using a microorganism of known resperation rate, R, which was separately determined by the Warburg manometric method. The method involved measuring the change in oxygen partial pressure in a flask filled with actual medium sampled from the bioreactor. The solubility S_0 is calculated as follows:

$$S_0 = R \Big/ \left(\frac{dp_0}{dt}\right) , \tag{67}$$

where R = respiration rate measured from Warburg method, $\left(\frac{dp_0}{dt}\right)$ = measured change in partial pressure.

When combined with enzymes, a DO probe becomes a sensor for specific substrates such as glucose, galactose, ethanol, methanol, etc. A number of enzymes, known as oxygen oxidoreductases, oxidizes substrates by utilizing molecular oxygen to form a product and H_2O_2. The formation of H_2O_2 can be monitored by the DO electrode, which is directly proportional to the substrate concentration[32]. Usually, an enzyme is entrapped or immobilized inside the membrane of the DO probe. Glucose electrodes commercially available are based on this principle. A family of polarographic enzyme electrodes for measuring alcohols and other substrates are described by Clark[32, 33].

8 Conclusions

Although it seems to be possible to design and construct DO probes to meet most of the requirements for specific applications, several problems still remain to be solved. Among them are: long term stability, calibration problems, and reliable measurement of low oxygen tension in dense microbial cultures. Some of the current trends and possible future developments are given below.

Better materials for the electrode components need to be continuously searched. A membrane material with antifouling feature is desirable to prevent growth of microorganisms on the surface. Heat resistance of the membrane need to be improved. The use of an ion-exchange membrane as the electrolyte[114] can certainly improve the probe performance. There has to be a solution to the aging of the cathode surface. Metal oxide may be an alternative. Zirconia cells used for oxygen measurement at high temperatures do not require calibration. Recently, β-alumina was shown to measure oxygen at room temperature[102]. The integrated circuit fabrication technique[137] can be further applied in making probes having uniform characteristics. The use of disposable sensing tips is currently employed by some manufacturers.

The method of operating DO probes can be improved. For example, in a long-term continuous measurement in bioreactors, the probe life can be extended if the probe is operated on a pulse mode under a microprocessor control. Since the probe will be on only for a fraction of time, the probe life can be extended and the probe malfunction can be found from the transient response each time when the probe is switched on. By improving the probe design and the instrumentation, it would be possible to have a long-lasting and self-calibrating probe system.

9 Acknowledgement

The authors wish to thank Mr. Alan Abel of Drexel University for his assistance in preparing this manuscript.

10 Nomenclature

a	gas-liquid interfacial area
A	surface area of cathode
A_1, A_2	quantities defined in Eq. (47)
B	quantity defined in Eq. (61)
B_1, B_2	thermistor constants defined in Eq. (48)
c	concentration of dissolved oxygen
c_g	inlet gaseous oxygen concentration
c'_g	oxygen concentration in dispersed gas phase
c_0	bulk concentration
c_1	initial concentration
c^*	equilibrium concentration
d_e	thickness of electrolyte layer
d_L	thickness of liquid film
d_m	thickness of membrane
$\bar{d}_t$	equivalent thickness defined by Eq. (20)
$\bar{d}$	equivalent thickness defined by Eq. (18)
d'	equivalent thickness defined by Eq. (53)
D_e	oxygen diffusivity of electrolyte layer
D_L	oxygen diffusivity of liquid film
D_m	oxygen diffusivity of membrane
D^*_m	oxygen diffusivity of membrane at base temperature
D_0	oxygen diffusivity of medium
D_1, D_2	oxygen diffusivities
E	activation energy for oxygen permeation
E_D	activation energy for oxygen diffusion
F	Faraday's constant (96,500 coulombs/g-equivalent)
G	gas flow rate
H	Henry's Law constant
H	heat of solution
I	oxygen current
I_G	probe current in gas phase
I_L	probe current in liquid phase
I_s	steady-state current
I_t	transient current
I_T	oxygen current as function of temperature
J	oxygen flux
k	probe constant defined by Eqs. (11) or (19)
k_e	mass transfer coefficient of electrolyte layer
k_L	liquid phase mass transfer coefficient
k_{LM}	mass transfer coefficient of liquid film around membrane
K_M	Michaelis-Menten constant
k_m	mass transfer coefficient of membrane
k_0	zeroth order reaction constant
K_0	overall mass transfer coefficient
L	quantity defined by Eq. (55)
N	number of electrons per mole of oxygen reduced
p	oxygen partial pressure
p_e	oxygen partial pressure at electrolyte/membrane interface
p_m	oxygen partial pressure at membrane/liquid interface
p'_m	oxygen partial pressure defined by Eq. (57)

p_0	oxygen partial pressure of bulk medium
P_e	oxygen permeability of electrolyte layer
P_L	oxygen permeability of liquid film
P_m	oxygen permeability of membrane
r	distance from center of cathode
r_0	radius of cathode
R	gas law constant or reaction rate
R_L	load resistance
R_T	resistance of thermistor
s	quantity defined by Eq. (29)
S_m	oxygen solubility of membrane
S_m^*	oxygen solubility of membrane at base temperature
S_0	oxygen solubility of medium
S_w	oxygen solubility of water
t	time
T	absolute temperature
v	liquid velocity
v_c	critical liquid velocity
V	voltage drop
V_g	volume of gas absorbed
V_G	volume of dispersed gas phase
V_L	volume of liquid
V_s	volume of absorbing solvent
V_T	voltage drop across thermistor
x	distance from the surface of cathode
a	Bunsen coefficient defined by Eq. (36)
μ_1, μ_2	liquid viscosities
$\Gamma_{95\%}$	95% response time
Γ	normalized probe response defined by Eq. (39)

11 References

1. Adams, A.P., Morgan-Hughes, J.O., Sykes, M.K.: Anesthesiology *22*, 575 (1967)
2. Aiba, S., Ohashi, M., Huang, S.Y.: Ind. Eng. Chem. Fundam. *7*, 497 (1968)
3. Aiba, S., Huang, S.Y.: J. Ferment. Technol. *47*, 372 (1969)
4. Aiba, S., Huang, S.Y.: Chem. Eng. Sci. *24*, 1149 (1969)
5. Aiba, S., Humphrey, A.E., Mills, N.: Biochemical Engineering, 2nd ed. New York: Academic Press 1973
6. Albanese, R.A.: J. Theor. Biol. *33*, 91 (1971)
7. ASTM D1589-60, Annual Book of ASTM Standards, Part 31, p. 438 (1978)
8. Bandyopadhay, B., Humphrey, A.E.: Biotechnol. Bioeng. *9*, 533 (1967)
9. Barr, R.E., Tang, T.E., Hahn, A.W.: Adv. Exp. Med. Biol. *94*, 17 (1978)
10. Battino, R., Clever, H.L.: Chemical Reviews *66*, 395 (1966)
11. Baumgärtl, H., Lübbers, D.W.: In: Oxygen supply. Kessler, M., Bruley, D.F., Clark, L.C., Lübbers, D.W., Silver, I.A., Strauss, J. (eds.), p. 130. London: University Park Press 1973
12. Baumgärtl, H., Grünewald, W., Lübbers, D.W.: Pflügers Arch. *347*, 49 (1974)
13. Beckman Technical Bulletine, Fieldlab Oxygen Analyzer, Beckman Instr. 1974
14. Beechey, R.B., Ribbons, D.W.: Methods in Microbiology, Vol. 6B, p. 25 (1972)
15. Benedeck, A.A., Heideger, W.J.: Water Res. *4* (9), 627 (1970)

16. Berkenbosch, A., Riedstra, J.W.: Acta Physiol. Pharmacol. Neerl. *12*, 144 (1963)
17. Berkenbosch, A.: Acta Physiol. Pharmacol. Neerl. *14*, 300 (1967)
18. Bicher, H.I., Knisely, M.H.: J. Appl. Physiol. *28*, 387 (1970)
19. Biomarine Industries, Tech. Bulletin: OM322-675, Malvern, Pennsylvania (1975)
20. Bird, B.D., Williams, J., Whitman, J.G.: Brit. J. Anaesth. *46*, 249 (1974)
21. Borkowski, J.D., Johnson, M.J.: Biotechnol. Bioeng. *9*, 635 (1967)
22. Bourdaud, D., Lane, A.G.: Biotechnol. Bioeng. *16*, 279 (1974)
23. Briggs, R., Viney, M.: J. Sci. Instr. *41*, 78 (1964)
24. Brookman, J.S.G.: Biotechnol. Bioeng. *11*, 323 (1969)
25. Brown, D.E.: In: Methods in microbiology. Norris, J.R., Ribbons, D.W. (eds.), Vol. 2, p. 125, 1970
26. Bühler, H., Ingold, W.: Process Biochem., p. 19, April (1976)
27. Bungay III., H.R., Huang, M.Y., Sanders, W.M.: Amer. Inst. Chem. Eng. J. *19*, 373 (1973)
28. Bungay III, H.R., Whalen, W.J., Sanders, W.M.: Biotechnol. Bioeng. *11*, 765 (1969)
29. Carey, F.G., Teal, J.M.: J. Appl. Physiol. *20*, 1074 (1965)
30. Carrit, D.E., Kanwisher, J.W.: Anal. Chem. *31*, 5 (1959)
31. Clark, L.C.: Trans. Am. Soc. Artif. Organs. *2*, 41 (1956)
32. Clark, L.C.: Biotech. Bioeng. Symp. *3*, 377 (1972)
33. Clark, L.C., Clark, E.W.: Adv. Exp. Med. Bio. *37*A, 127 (1973)
34. Cobbold, R.S.C.: Transducers for biomedical measurements. New York: John Wiley and Sons 1974
35. Danckwerts, P.V.: Gas-liquid reactions, p. 18. New York: McGraw Hill 1970
36. Dang, N.D.P., Karrer, D.A., Dunn, I.J.: Biotechnol. Bioeng. *19*, 853 (1977)
37. Davies, P.W.: In: Physical techniques in biological research *4*, 137 (1961)
38. Davies, P.W., Brink, F., Jr.: Rev. Sci. Instr. *13*, 524 (1942)
39. Döhring, W., Diefenthaler, K., Barnikol, W.K.: In: Oxygen supply. Kessler, M., Bruley, D.F., Clark, L.C., Lübbers, D.W., Silver, I.A., Strauss, J. (eds.), p. 86. London: University Park Press 1973
40. Dunn, I.J., Einsele, A.: J. Appl. Chem. Biotechnol. *25*, 707 (1975)
41. Eberhard, P., Mindt, W., Jann, F., Hammacher, K.: Med. Biol. Eng. *13*, 436 (1975)
42. Elsworth, R.: Brit. Chem. Eng., p. 63, Feb. (1972)
43. Evangelista, R., Gorin, N.H., Hamilton, W.J.: Power *121*, 83 (1977)
44. Evans, D.H., Lingane, J.J.: Electroanal. Chem. *6*, 283 (1963)
45. Fatt, I.: J. Appl. Physiol. *19*, 326 (1964)
46. Fatt, I.: Anal. N. Y. Acad. Sci. *148*, 81 (1968)
47. Fatt, I.: Polarographic oxygen sensors. Ohio: CRC Press 1976
48. Fatt, I., St. Helen, R.: J. Appl. Physiol. *27*, 435 (1969)
49. Flynn, D.S., Kilburn, D.G., Lilly, M.D., Webb, F.C.: Biotechnol. Bioeng. *9*, 623 (1967)
50. Forbes, M., Lynn, S.: Amer. Inst. Chem. Eng. J. *21*, 763 (1975)
51. Friesen, W.O., McIlory, M.B.: J. Appl. Physiol. *29*, 258 (1970)
52. Gal-Or, B., Hauk, J.P., Hoelscher, H.E.: Int. J. Heat. Mass Transfer *10* 1559 (1967)
53. Gelder, R.L., Neville, J.F.: Amer. J. Clin. Path. *55*, 325 (1971)
54. Gobe, J.H., Phillips, T.: Biotechnol. Bioeng. *6*, 491 (1964)
55. Goldstick, T.K., Fatt, I.: Chem. Eng. Prog. Symp. Ser. *66*, 101 (1970)
56. Grünewald, W.: Pflügers Arch. *320*, 24 (1970)
57. Grünewald, W.: Pflügers Arch. *322*, 109 (1971)
58. Grünewald, W.: In: Oxygen supply. Kessler, M., Bruley, D.F., Clark, L.C., Lübbers, D.W., Silver, I.A., Strauss, J. (eds.), p. 160. London: University Park Press 1973
59. Harrison, D.E.F., Melbourne, K.V.: Biotechnol. Bioeng. *12*, 633 (1970)
60. Harrison, D.E.F.: J. Appl. Chem. Biotech. *22*, 417 (1972)
61. Harrison, D.E.F.: In: Measurement of oxygen. Degn, H., Balslev, I., Brook, R. (eds.). New York: Elsevier Scientific Publishing Co. 1976
62. Heineken, F.G.: Biotechnol. Bioeng. *12*, 145 (1970)

63. Heineken, F.G.: Biotechnol. Bioeng. *8*, 599 (1971)
64. Hitchman, M.L.: Measurement of dissolved oxygen. New York: John Wiley and Sons 1978
65. Hospodka, J., Caslavsky, Z.: Folia Microbiol. (Prague) *10*, 186 (1965)
66. Huang, M.Y., Bungay III., H.R.: Biotechnol. Bioeng. *15*, 1193 (1973)
67. Hulands, G.H., Nunn, J.F., Paterson, G.M.: Brit. J. Anaesth. *42*, 9 (1970)
68. Hwang, S.T., Choi, C.K., Kammermeyer, K.: Sep. Sci. *9*, 461 (1974)
69. Jensen, O.J., Jacobsen, T., Thomsen, K.: J. Electroanal. Chem. *87*, 203 (1978)
70. Jeter, H.W., Foyn, E., King, M., Gordon, L.I.: Limnol. Oceanog. *17*, 288 (1972)
71. Johnson, M.J., Borkowski, J., Engblom, C.: Biotech. Bioeng. *6*, 457 (1964)
72. Jones, C., Pust, H., Lauer, J.: Analysis Instrumentation *14*, 51 (1976)
73. Kessler, M.: In: Oxygen supply. Kessler, M., Bruley, D.F., Clark, L.C., Lübbers, D.W., Silver, I.A., Strauss, J. (eds.), p. 81. London: University Park Press 1973
74. Key, A., Parker, D., Davies, R.: Phys. Med. Biol. *15*, 569 (1970)
75. Kimmich, H.P., Kreuzer, F.K.: Prog. Resp. Res. *3*, 100 (1969)
76. Kinsey, D.W., Bottomley, R.A.: J. Inst. Brewing *69*, 164 (1963)
77. Kjaergaard, L.: Biotechnol. Bioeng. *18*, 729 (1976)
78. Kliner, A.A.: Brit. Chem. Eng. *10*, 537 (1965)
79. Kok, R., Zajic, J.E.: Can. J. Chem. Eng. *51*, 782 (1973)
80. Kok, R., Zajic, J.E.: Biotechnol. Bioeng. *17*, 527 (1975)
81. Kok, R.: Biotechnol. Bioeng. *18*, 729 (1976)
82. Kolthoff, I.M., Lingane, J.J.: Polarography. New York: Interscience 1952
83. Krebs, W.M., Haddad, I.A.: In: Developments in industrial microbiology, Vol. 13. Washington D.C.: Chap. 11. Am. Inst. Biol. Sci. 1972
84. Kreuzer, F., Kimmich, H.P.: In: Measurement of oxygen. Degn, H., Balslev, I., Brook, R. (eds.), p. 123. New York: Elsevier Sci. Pub. Co. 1976
85. LaForce, R.C.: In: Polarographic oxygen sensors. Fatt, I. (ed.). Ohio: CRC Press 1976
86. Lee, Y.H.: Ph. D. Thesis. Purdue University 1977
87. Lee, Y.H., Tsao, G.T., Wankat, P.C.: Ind. Eng. Chem. Fundam. *17*, 59 (1978)
88. LeFevre, M.E.: J. Appl. Physiol. *26*, 844 (1969)
89. LeFevre, M.E., Wyssbrad, H.R., Brodsky, W.A.: Bioscience *20*, 761 (1970)
90. Linek, V.: Biotechnol. Bioeng. *14*, 285 (1972)
91. Linek, V., Sobotka, M., Prokop, A.: Biotech. Bioeng. Symp. *4*, 429 (1973)
92. Linek, V., Vacek, V.: Biotechnol. Bioeng. *18*, 1537 (1976)
93. Linek, V., Benes, P.: Biotechnol. Bioeng. *19*, 741 (1977)
94. Linek, V., Vacek, V.: Biotechnol. Bioeng. *19*, 983 (1977)
95. Linek, V., Vacek, V.: Biotechnol. Bioeng. *20*, 305 (1978)
96. Linek, V., Benes, P.: Biotechnol. Bioeng. *20*, 903 (1978)
97. Lipner, H., Witherspoon, L.R., Champeaux, V.C.: Anal. Chem. *36*, 204 (1964)
98. Liu, M.S., Branion, R.M.R., Duncan, D.W.: Biotechnol. Bioeng. *15*, 213 (1973)
99. Lloyd, B.B., Seaton, B.: J. Physiol. (London) *207*, 29 (1970)
100. Lübbers, D.W.: Progr. Resp. Res. *3*, 112 (1969)
101. Lübbers, D.W., Baumgärtl, H., Fabel, H., Huch, A., Kessler, M., Kunze, K.: Progr. Resp. Res. *3*, 136 (1969)
102. Lundsgaard, J.S.: In: Measurement of oxygen. Degn, H., Balslev, J., Brook, R. (eds.), p. 159. New York: Elsevier Scientific Pub. Co. 1976
103. Lundsgaard, J.S., Gronlund, J., Degn, H.: Biotechnol. Bioeng. *20*, 809 (1978)
104. Mackereth, F.J.H.: Rev. Sci. Instr. *41*, 38 (1964)
105. Mancy, K.H., Westgarth, W.C.: J. Water Pollut. Control *34*, 1037 (1962)
106. Mancy, K.H., Okun, D.A., Reilley, C.N.: J. Electroanal. Chem. *4*, 65 (1962)
107. Mclennan, D.G., Pirt, S.J.: J. Gen. Microbiol. *45*, 289 (1966)
108. Mertu, K., Dunn, I.J.: Biotechnol. Bioeng. *18*, 591 (1976)
109. Misra, H.P., Fridovich, I.: Anal. Biochem. *70*, 632 (1976)
110. Modern Plastics Encyclopedia *51* (10A), 730 (1974–1975)

111. Moran, F., Kettel, L.J., Cugell, D.W.: J. Appl. Physiol. *21*, 725 (1966)
112. Morisi, G., Gualandi, G.: Biotechnol. Bioeng. *8*, 625 (1965)
113. Newley, R., Macpherson, P.B.: Wear *45*, 395 (1977)
114. Niedrach, L.W., Stoddard, W.H.: U.S. Patent 3,703,457 (1972)
115. Nikolaev, P.I., Polyanskii, V.P., Kantere, V.M., Kharitonova, E.V.: Theoretical foundations of Chem. Eng. (Engl. Translation) *10*, 13 (1976)
116. Ohashi, M., Watabe, T.: Paper presented in 2nd International Conf. on Computer Applications in Fermentation Tech., August 28–30, U. of Penn., U.S.A. 1978
117. Orbispheres Tech. Bull., Orbisphere Lab., York, Maine (1976)
118. Phillips, D.H., Johnson, M.J.: Biochem. Microbiol. Technol. Eng. *3*, 261 (1961)
119. Pittman, R.W.: Nature *195*, 449 (1962)
120. Radlett, P.J., Breame, A.J., Telling, R.C.: Lab. Pract. *21*, 81 (1972)
121. Parker, D.H., Clifton, J.S.: Phys. Med. Biol. *12*, 193 (1967)
122. Reeves, J.R.: Water & Sewage Works, Feb 44 (1976)
123. Rexnord Technical Bulletine 773, Rexnord ppb DO analyzer, Rexnord Co., Malvern, Pennsylvania 1977
124. Robinson, C.W., Wilke, C.R.: Biotechnol. Bioeng. *15*, 755 (1973)
125. Roughton, F.J.W., Scholander, P.F.: J. Biol. Chem. *148*, 541 (1943)
126. Rowley, B.I.: In: Automation, mechanization, and data handling in microbiology. Baillie, A., Gilbert, R.J. (eds.), p. 163. New York: Acad. Press 1970
127. Sawyer, D.T., Interrante, L.V.: J. Electroanal. Chem. *2*, 310 (1961)
128. Schuler, R., Kreuzer, F.: Progr. Resp. Res. *3*, 64 (1969)
129. Schuchhardt, S., Lösse, B.: In: Oxygen supply. Kessler, M., Bruley, D.F., Clark, L.C., Lübbers, D.W., Silver, I.A., Strauss, J. (eds.), p. 108. London: University Press 1973
130. Severinghaus, J.W.: Clin. Chem. *9*, 727 (1963)
131. Severinghaus, J.W.: Annal. N.Y. Acad. Sci. *148*, 115 (1968)
132. Siedell, A., Linke, W.F.: Solubilities of inorganic and metal organic compounds, 4th ed., Vol. 2. Amer. Chem. Soc. 1965
133. Silver, I.: Med. Electron. Biol. Eng. *1*, 547 (1963)
134. Silver, I.A.: Med. Elect. Biol. Eng. *3*, 377 (1965)
135. Silver, I.A.: Adv. in Chem. Ser. *118*, 343 (1973)
136. Silver, I.A.: Adv. Expt. Med. Biol. *37*A, 7 (1974)
137. Siu, W., Cobbold, R.S.C.: Med. Biol. Eng. *14*, 109 (1976)
138. Stuck, J.D., Howell, J.A., Cullinan, H.T., Jr.: J. Theor. Biol. *31*, 509 (1971)
139. Tsao, G.T., Lee, D.D.: Amer. INst. Chem. Eng. J. *21*, 979 (1975)
140. Tsao, G.T., Lee, Y.H.: Ann. Reports in Ferm. Proc. *1*, 115 (1977)
141. Tuffile, C.M., Pinho, F.: Biotechnol. Bioeng. *12*, 849 (1970)
142. Vincent, A.: Process Biochem. April, 19 (1974)
143. Votruba, J., Sobotka, M.: Biotechnol. Bioeng. *18*, 1815 (1976)
144. Votruba, J., Sobotka, M., Prokop, A.: Biotechnol. Bioeng. *19*, 435 (1977)
145. Votruba, J., Sobotka, M., Prokop, A.: Biotechnol. Bioeng. *20*, 913 (1978)
146. Wennberg, L.A.: J. Appl. Physiol. *38*, 540 (1975)
147. Wernau, W.C., Wilke, C.R.: Biotechnol. Bioeng. *15*, 571 (1973)
148. Whalen, W.J., Riley, J., Nair, P.: J. Appl. Physiol. *23*, 798 (1967)
149. Whalen, W.J.: Progr. Resp. Res. *3*, 158 (1969)
150. Wilke, C.R., Chang, P.: Amer. Inst. Chem. Eng. J. *1*, 264 (1955)

Power Consumption in Aerated Stirred Tank Reactor Systems

Heinz Brauer
Institut für Chemieingenieurtechnik
Technische Universität Berlin, D-1000 Berlin, West Germany

1 Introduction . . . 87
2 Fundamentals of Energy Transfer and Gas Dispersion . . . 88
2.1 Functions of Stirrers . . . 88
2.2 Arrangement of Stirrers . . . 90
2.3 Fluid and Bubble Movement . . . 92
2.4 Bubble Generation . . . 94
2.5 Energy Transfer Curve . . . 94
3 Energy Transfer in Pure Liquids . . . 100
3.1 Energy Transfer in Pure Newtonian Liquids . . . 100
3.1.1 Energy Transfer by Turbine and Paddle Stirrers . . . 100
3.1.2 Energy Transfer by Propeller and Disk Stirrers . . . 111
3.2 Energy Transfer in Pure Non-Newtonian Liquids . . . 113
3.2.1 Energy Transfer Calculation Method after Metzner and Otto . . . 113
3.2.2 Energy Transfer Calculation Method after Schilo . . . 114
4 Energy Transfer in Aerated Systems . . . 115
5 Symbols . . . 117
6 References . . . 118

The aim of this contribution is to give all biochemists and biochemical engineers working in the field of biotechnology an insight into the information which is available on energy transfer in mixing equipment under various operational conditions.

1 Introduction

The aeration of cultivation media is a process that is not yet well understood since comprehensive studies are lacking. This is primarily because the properties of such fluids cannot be easily described[1) consequently, aeration studies of growth media are usually carried out with the real fluids under actual operational conditions. The results so obtained are of limited utility. Except where substantial progress has been achieved in the characterization of this type of fluids, all investigations on the fluid dynamic behavior and mass and heat transfer processes within these fluids will be restricted to special cases without general applicability.

One of their important properties is the viscosity. The magnitude of this property ranges over several powers of ten, when various types of fluids are considered. But the situation becomes even worse when viscosity is described as a function of batch process time since virtually all the process properties vary over the time course. The consequence is that the aeration process, power consumption and gas contact will vary during a batch process.

Culture media generally show non-Newtonian behavior. The viscosity of such fluids not only depends on temperature and the concentration of the various components but also on the shear stress that is related to the movement of the fluid. The stress dependency of viscosity is not a general function for all non-Newtonian fluids. There are many groups of reaction fluids, each of which has a characteristic stress dependency for viscosity and its stress dependency has to be determined for each fluid.

The typical bioreactor is a squat cylindrical vessel with a stirrer that keeps the fluid in motion and disperses gas (Fig. 1). The fluid motion varies widely within the vessel and the shear stress and consequently the viscosity vary locally.

The local properties of the fluids exert a considerable influence on the movement of the gas bubbles that have been generated by the stirrer. These local properties are unknown and they will probably remain unknown because of the very complex three dimensional fluid flow in the conventional type of vessel. Therefore no sound physical basis for the description of bubble movement within a vessel is available. Recourse has to be taken either to purely empirical or to qualitative descriptions of bubble movement.

So far as power consumption is concerned, the situation is much better. The transfer of mechanical energy from the stirrer to the surrounding reaction fluid and to the gas that is to be dispersed takes place at the surface of the stirrer. It has been found, at least for a large group of non-Newtonian fluids, that a mean value of the viscosity of the fluid in contact with the stirrer surface can be defined and used for the calculation of energy transfer or power consumption.

2 Fundamentals of Energy Transfer and Gas Dispersion

2.1 Functions of Stirrers

The shape and the arrangement of a stirrer within a vessel should reflect the functions which the stirrer has to meet. These functions are as follows:

(a) Energy transfer to the fluid,
(b) Dispersion of gas in the liquid,
(c) Separation of gas and liquid,
(d) Mixing of all components of the process fluid.

Therefore the stirrer is at the same time an energy transfer element, a gas dispersion element, a separation element, and a mixing element.

The energy transferred by the stirrer to the fluid is required to achieve suitable fluid movement in the vessel. Fluid movement is always connected with energy dissipation. Mechanical energy is thereby converted into thermal energy. The "loss" of mechanical energy must be balanced by mechanical energy constantly transferred to the fluid by

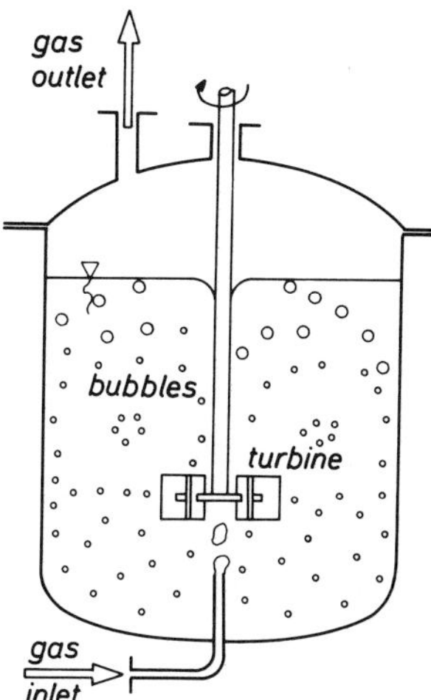

Fig. 1. General layout of a simple reactor

the stirrer. The energy is necessary for a particular kind of fluid movement that is best suited for the process. This fluid movement should be achieved by a minimum of energy. The function of a stirrer as an energy transfer element may be described as follows:

To achieve a prescribed fluid movement within a vessel with a minimum of energy.

Energy is not only required for fluid movement but for gas bubble generation too. However this energy is a very small fraction of the overall energy transferred and may be neglected. The gas dispersion process consists of two parts:

(a) Gas bubble generation,

(b) Gas bubble distribution in the fluid.

The second part of the process depends primarily on the fluid movement achieved by the stirrer. The all important function of a stirrer as element for gas dispersion is the generation of gas bubbles. The mass flux across the interface between a gas and a liquid is proportional to the interfacial area. Increased interfacial area will lead to an increased mass flux. A large interfacial area requires the generation of small bubbles from a given gas flow rate and optimum mass transfer conditions may require a prescribed bubble diameter. The function of a stirrer as a gas dispersion element may be described as follows:

To achieve a prescribed interfacial area between gas and liquid and its distribution in the vessel.

In many cases it may happen that surface active agents concentrate in the gas/liquid interface thereby hindering the diffusional process across the interface. Generation of interfacial area must therefore include continuous renewal.

The third function of a stirrer is concerned with the separation of gas and liquid. The separation process is often more complicated than the dispersion process. Big bubbles separate easily from the liquid, while the separation of very small bubbles may be extremely difficult. The separation process depends on the other hand heavily on the fluid flow in the vessel and the shape of the vessel. The function of a stirrer as a separation element may be described as follows:

To achieve such conditions for bubble diameter and bubble movement in the vessel that separation of gas and liquid will result easily.

The fourth function of a stirrer is generally considered as the most important. All components of the biological suspension should be completely mixed if possible. This includes the components in the liquid, the microorganisms and the gas bubbles. When all components are mixed in an ideal way the consequence is a constant rate of reaction in all volumetric elements of the biological suspension and the concentrations and temperature will be the same throughout. But this is never the case because the achievement of ideal mixing depends on the fluid flow in the vessel, and requires uneconomic energy transfer. Research furthermore has established that in vessels with conventional rotating mixing elements the quality of the mixture is restricted by the imposed rotational movement of the fluid[2)]. After these critical remarks the function of a stirrer as a mixing element may be described as follows:

To achieve a state of mixture that is favorable to the rate of reaction in the biological suspension with a minimum of energy and without inflicting any harm on the microorganisms.

Most of the stirrers used in technical equipment have not been designed on the basis of the functions described. This is because the understanding of the extremely complicated phenomena of stirred gas/liquid systems has become available only recently.

2.2 Arrangement of Stirrers

Although a great number of different types of stirrer are available only a few have gained importance and become widely used. These stirrers are as follows:

1. Turbine stirrer
2. Paddle stirrer
3. Propeller stirrer
4. Disk stirrer.

The disk stirrer is not yet widely used. But it has some exceptional properties that may increase its importance.

The stirrers are shown in Figs. 2–5 with the dimensions identified as follows:

d_r Stirrer diameter
h_r Stirrer height, blade height, paddle height
d_s Plate diameter of turbine stirrer
h_s Plate thickness of turbine stirrer
h_p Pitch of propeller
l Length of blade and paddle
b Thickness of blade and paddle
d_L Hole diameter of disk stirrer
d_K Centre line diameter of disk stirrer
c Number of blades of turbine, paddle, and propeller stirrer
z Number of holes and disk stirrer.

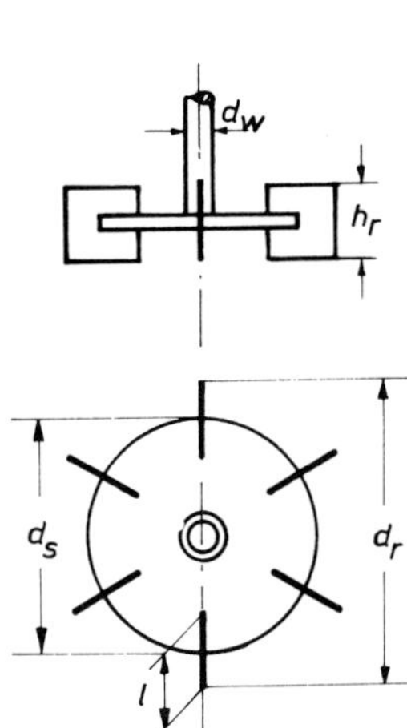

Fig. 2. Turbine stirrer

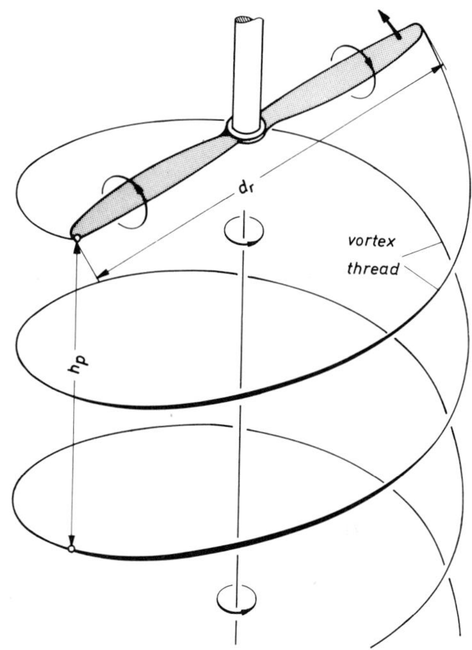

Fig. 4. Propeller stirrer

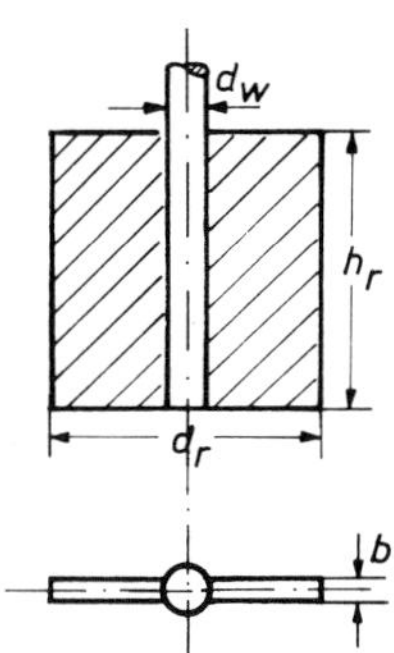

Fig. 3. Paddle stirrer

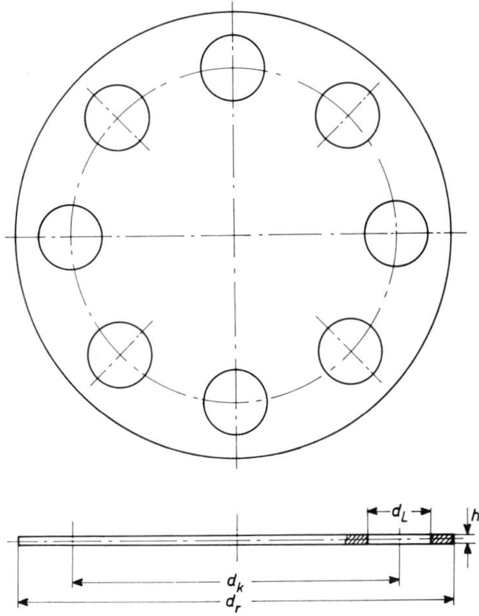

Fig. 5. Disk stirrer

The arrangement of a stirrer in a vessel is shown in Fig. 6. The vessels may or may not be equipped with baffles. The important dimensions are the following:

D Vessel diameter
H Liquid height in vessel
e Distance from bottom of vessel
h_w Height of rotating axis in liquid
s Radial length of baffle.

When arranging stirrers and baffles in vessels the following values of the ratios of the geometrical dimensions are normally observed:

$H/D \approx 1$	For all types of stirrers
$s/D \approx 1/10$	For turbine and paddle stirrers
$e/D \approx 1/3$	For all types of stirrers
$d_r/D \approx 1/3$	For turbine and paddle stirrers
$d_r/D \approx 1/2-2/3$	For propeller and disk stirrers.

2.3 Fluid and Bubble Movement

The movement of the fluid in a mixing vessel enforced by a rotating stirrer may be divided into primary and secondary flow patterns. The primary movement is the rotational or tangential flow of the fluid as shown in Fig. 7. Most of the energy transferred from the stirrer to the fluid is consumed by this movement. So far as the mixing process is concerned, the primary movement is of minor importance and in this respect the energy contained in the primary movement is wasted.

The secondary movement is shown in Fig. 8. This consists of radial and axial components. In mixing processes the secondary movement is the most important part of the flow pattern. Only a small fraction of the energy transferred to the fluid is contained in this secondary movement. The secondary flow is closely related to regions of very slow movement. These latter regions are to be found in the corners of the flow field and in vortex centers; they are badly mixed.

The strength of the secondary movement depends on the centrifugal forces induced by the primary rotational fluid movement. This movement is strongest close to the rotating stirrer and consequently so is the centrifugal force. Close to the surface of the stirrer a fluid movement takes place in the radial outward direction. The strength of this fluid motion depends on the shape of the stirrer which should be such that the fluid movement in the radial outward direction is opposed by the lowest possible resistance force. This is the case with the disk stirrer.

Primary and secondary fluid movement occur in both single- and multi-phase systems. But with increasing gas flow rate the influence of the stirrer on the fluid motion decreases and the gas flow determines the main movement[3)]. Figure 9 shows the movement of the gas bubbles at low gas flow rate, the bubble movement is determined by the secondary flow of the liquid that is enforced by the stirrer. At very high gas flow rates the bubble movement enforces a liquid flow as shown in Fig. 10, i.e., the stirrer has little influence. When normal operational conditions prevail in a bioreactor the bubble movement is still determined by the liquid movement, i.e., as Fig. 9. Primary and secondary movement

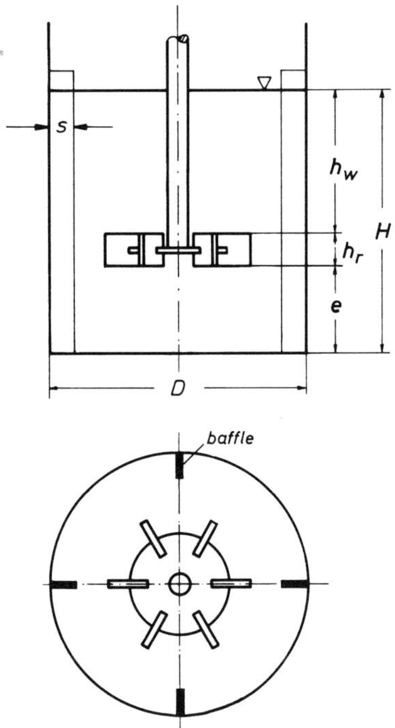

Fig. 6. Arrangement of a stirrer in a vessel

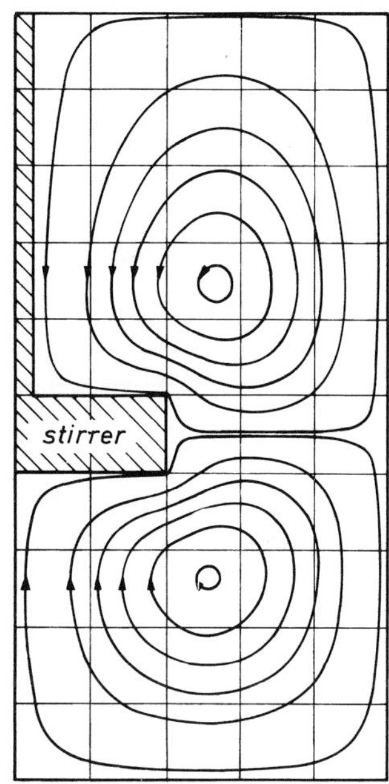

Fig. 8. Streamlines of secondary fluid movement

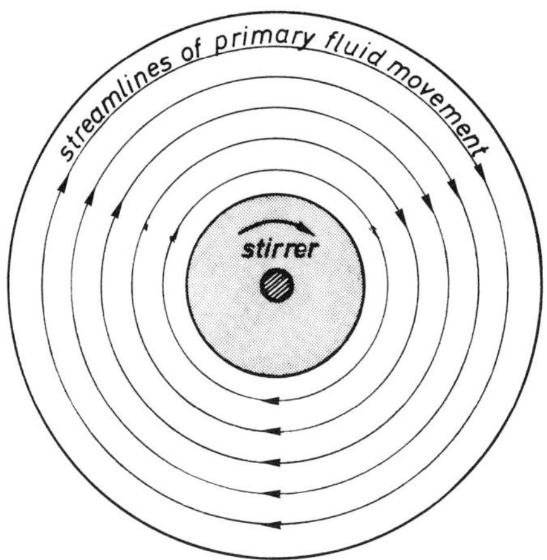

Fig. 7. Streamlines of primary fluid movement

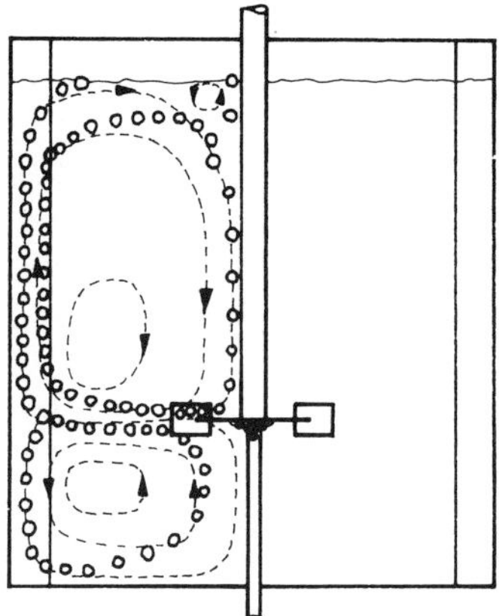

Fig. 9. Movement of gas bubbles in a vessel at low gas flow rate

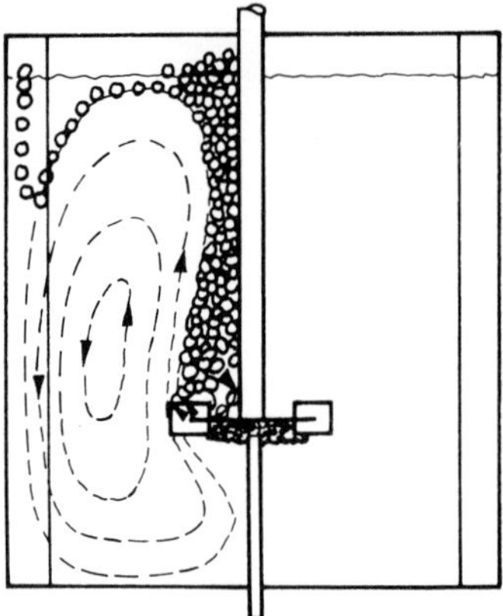

Fig. 10. Movement of gas bubbles in a vessel at high gas flow rates

are of equal importance. The rotational primary movement will carry the bubbles along and increase their residence time in the vessel. In turn this will increase the gas hold-up which will increase the turbulence in the liquid. A high degree of turbulence will both enhance the mass transfer to the liquid phase and aid liquid phase mixing.

2.4 Bubble Generation

Bubble generation has been carefully investigated in the last few years. Although the methods applied have only involved visual and photographic observations, a good understanding of bubble generation has been achieved. However quantitative description of bubble generation is not yet possible. The fundamental work has been carried out by Biesecker[3)] and van't Riet[4)].

Bubble generation by turbine stirrers is shown in Figs. 11 and 12, by propeller stirrers in Figs. 13 and 14, and by disk stirrers in Fig. 15. According to these figures bubble generation takes place in regions of reduced pressure. Such regions exist in the flow field behind obstacles, for instance behind the blades of turbine and propeller stirrers where vortex threads or vortex sheets may be observed. At low gas flow rates bubble generation takes place in vortex threads (Figs. 11 and 13). At higher gas flow rates the generation of bubbles occurs in vortex sheets thereby increasing the rate of bubble generation (Figs. 12, 14, and 15). These figures prove that dispersion elements should be designed in such a way that they produce suitable regions of reduced pressure. Bubble generation for paddle stirrers is the same as described for turbine stirrers. For turbine stirrers bubble generation has been sketched for a single-blade stirrer, but the effect is the same for four-, six- or even eight-blade stirrers.

2.5 Energy Transfer Curve

The energy transfer curve, or power curve, gives the relation between the energy transfer coefficient and the relevant physical and fluid parameters. The energy transfer is

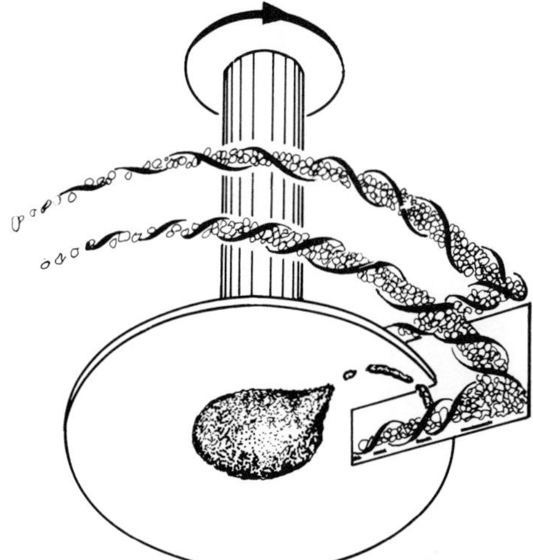

Fig. 11. Bubble generation in vortex threads in the wake of a turbine stirrer

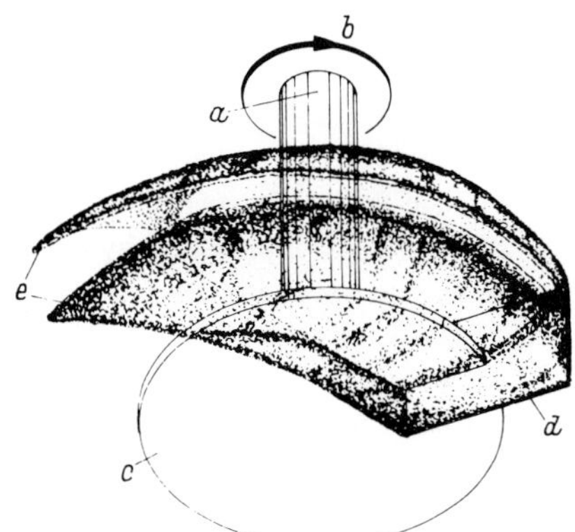

Fig. 12. Bubble generation in vortex sheets in the wake of a turbine stirrer

expressed as the Power- or Newton number:

$$\mathrm{Ne} \equiv \frac{\mathrm{N}}{\mathrm{n}^3 \mathrm{d}_\mathrm{r}^5 \varrho}, \tag{1}$$

where N is the energy transferred by the stirrer to the fluid, n is the number of revolutions of the stirrer per unit time, d_r is the diameter of the stirrer, and ϱ the density of liquid. In most cases the Newton number is presented as a function of the Reynolds number:

$$\mathrm{Re} \equiv \frac{\mathrm{n}\, \mathrm{d}_\mathrm{r}^2 \varrho}{\eta}, \tag{2}$$

where η is the viscosity of a Newtonian fluid.

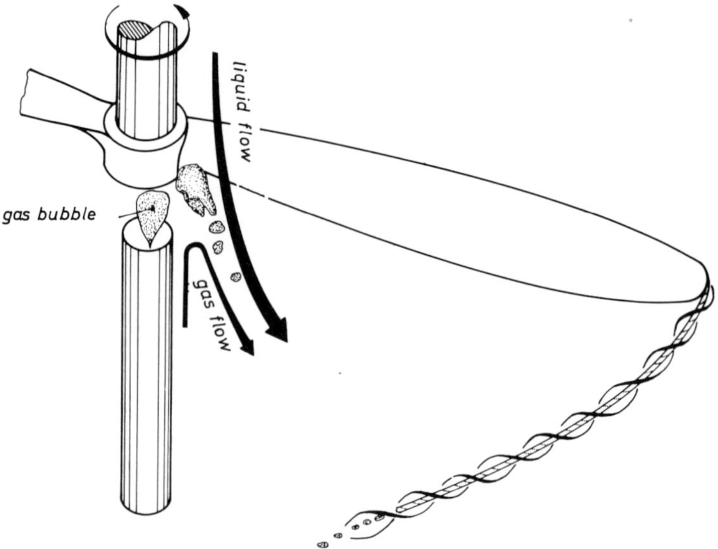

Fig. 13. Bubble generation in vortex threads in the wake of a propeller stirrer

There is no method available that permits the theoretical calculation of the Newton number, and it is not anticipated that such a method will be available in the foreseeable future. The functional relationship between the Newton number and the Reynolds number has to be determined by experiment.

The data given in Fig. 16 were obtained with a six-bladed turbine stirrer in a vessel with no baffles. Curve (*a*) fits the data closely. At values of the Reynolds number below 10 the Newton number Ne is proportional to Re^{-1}. With increasing values of Re the slope of the curve changes. Beyond $Re \approx 10^2$ the Newton number is proportional to $Re^{-0.28}$. According to the experimental evidence available the energy transfer curve consists of three distinct parts that reflect well defined flow regions:

(a) laminar flow.
  Reynolds number range: $0 \leqslant Re \lesssim 10$
  Ne-Re relationship: $Ne \sim Re^{-1}$.

(b) transition from laminar to turbulent flow.
  Reynolds number range: $10^1 \lesssim Re \lesssim 10^2$
  Ne-Re relationship: changing from $Ne \sim Re^{-1}$ to $Ne \sim Re^{-0.28}$.

(c) turbulent flow.
  Reynolds number range: $10^2 \lesssim Re \leqslant \infty$
  Ne-Re relationship: $Ne \sim Re^{-0.28}$.

For different stirrers the transition region may shift slightly to either lower or higher values of the Reynolds number and the exponent of the Reynolds number in the Ne–Re relationship may change slightly too.

A close inspection of the experimental data presented in Fig. 16 reveals that there are 11 groups of data. Each group of data covers a relatively small range of the Reynolds number which has a well defined upper limit given by Re_1. When the Reynolds number

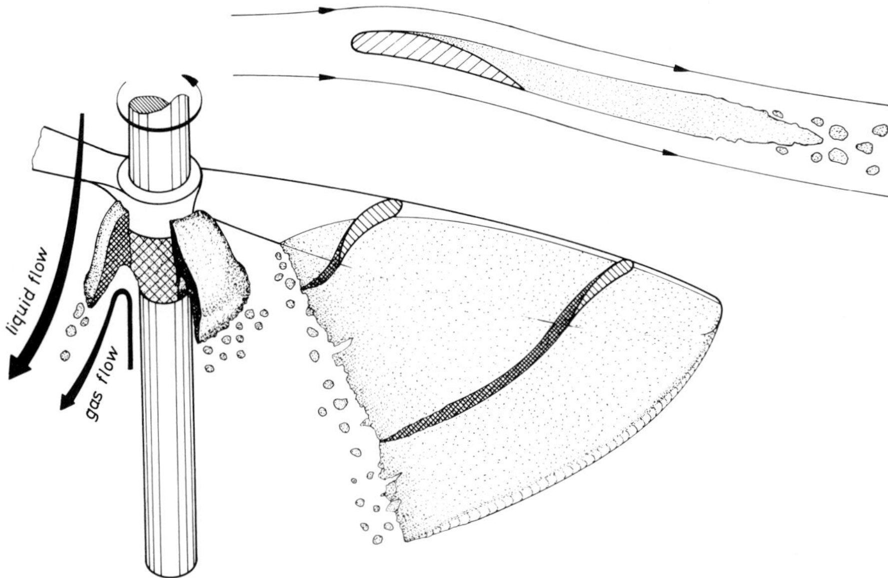

Fig. 14. Bubble generation in vortex sheets in the wake of a propeller stirrer

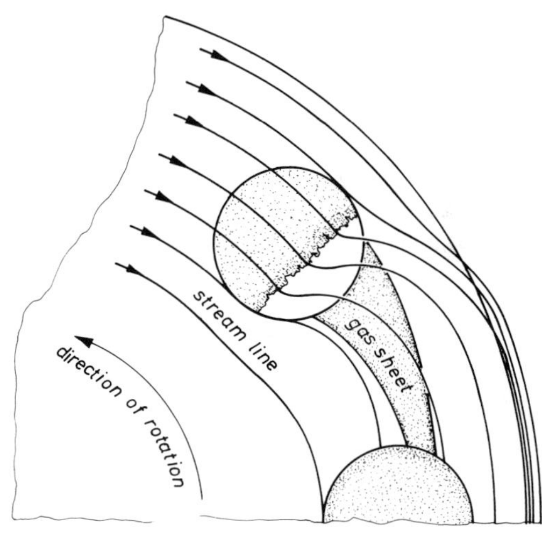

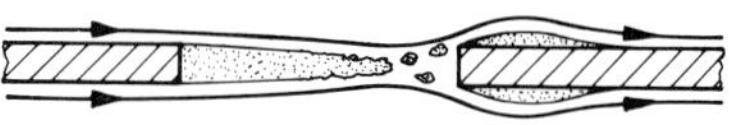

Fig. 15. Bubble generation in gas sheet of a disk stirrer

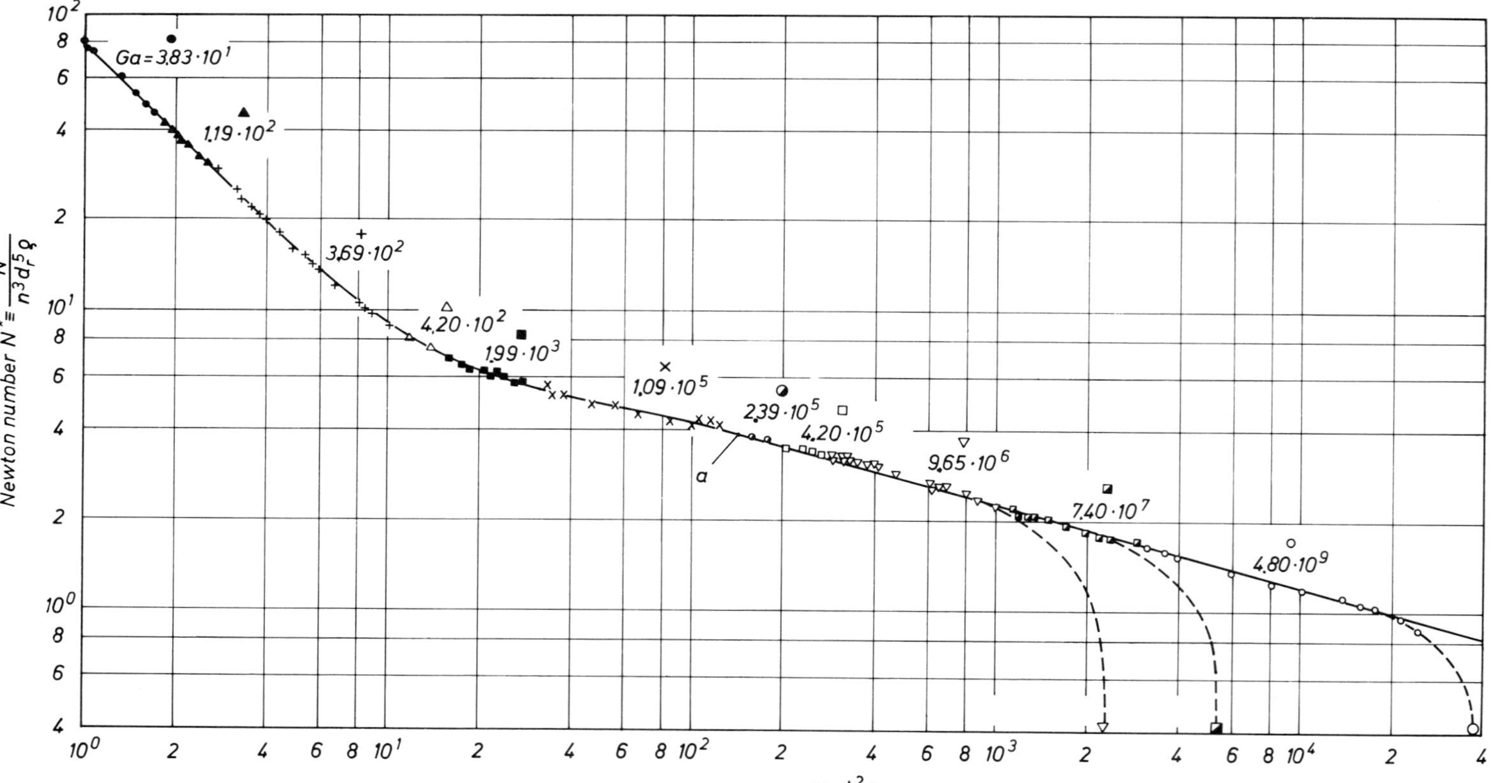

Fig. 16. Energy transfer curve for a turbine stirrer in a vessel without baffles

exceeds this limiting value operation of the stirrer becomes unstable. The transition from stable to unstable operation starts with a strong curl like deformation of the fluid surface around the rotating stirrer shaft. When the curl reaches the stirrer the energy is transferred intermittently. At the same time the energy transfer to the fluid is reduced and gas is dispersed in the liquid. Some of the data, given by the broken lines in Fig. 16 show this situation. It is advisable to stay well below the upper limiting Reynolds number Re_1 for stable operation.

The lower limiting value of the Reynolds number is clearly zero, however experimental limitations usually result in a lower limit about $Re_1/3$. This value depends on the sensitivity of the instrument that measures the torque in the rotating shaft of the stirrer and has no relationship with any basic fluid mechanical considerations. The data were obtained in the same experimental unit with different viscosity fluids. Theoretical investigations[5–7] disclosed that the transition to unstable operation conditions is a function of the Galileo number:

$$\mathrm{Ga} \equiv \frac{d_r^3 \, g \, \varrho^2}{\eta^2} \, , \tag{3}$$

where g is the gravitational acceleration. The upper limiting value of the Reynolds number is a function of the Galileo number and other parameters. In general the Newton number is a function of Re and Ga and further geometrical parameters:

$$\mathrm{Ne} = f\,(\mathrm{Re}; \mathrm{Ga}; \text{geometrical parameters}) \, . \tag{4}$$

This equation holds for pure fluids only. In the case of liquid/gas systems some further parameters must be introduced, taking into account the gas flow rate and gas properties.

The data for energy transfer in Fig. 16 were obtained in vessels without baffles. In Fig. 17 data are presented for vessels with and without baffles. The influence of the baffles on the Newton number is observed in the turbulent region only. At a Reynolds number of 10^4 the Newton number for vessels with baffles is roughly three times that for vessels without baffles. It is therefore advisable to use vessels without baffles if possible.

Baffles are quite often assumed to prevent the formation of the curl-like deformation of the fluid surface, but careful experiments have shown this to be wrong[8]. In vessels with baffles the curl-like deformation of the fluid surface is observed also although in quite a different way. The life-time of the curl-like deformation is limited in the presence of baffles and is unlimited in the absence of baffles. When baffles are present the surface deformations move about the vessel while they stay in a fixed position around the rotating stirrer shaft in vessels without baffles.

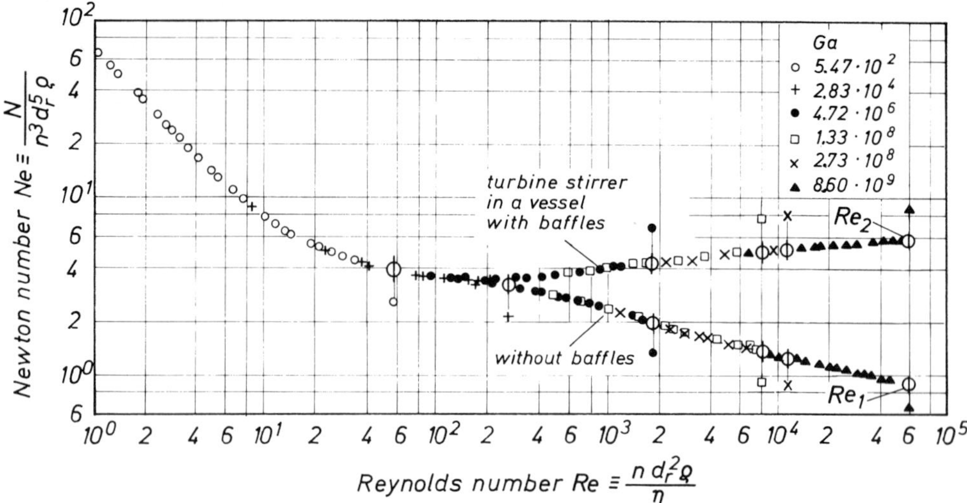

Fig. 17. Energy transfer curve for a turbine stirrer in vessels with and without baffles

3 Energy Transfer in Pure Liquids

The discussion starts with the case of energy transfer in Newtonian liquids. In the second section of this chapter it is explained how the information available for Newtonian fluids may be used to calculate the energy transfer in simple non-Newtonian fluids.

3.1 Energy Transfer in Pure Newtonian Liquids

Equations and graphs will be presented which permit the calculation of energy transfer to pure Newtonian liquids by turbine, paddle, propeller, and disk stirrers. For turbine and paddle stirrers the energy transfer process has been successfully analyzed. A model for energy transfer has been developed on the basis of which a theoretical treatment of the transfer process proved to be possible. The results are equations for energy transfer. In the case of propeller and disk stirrers comparable results for energy transfer calculations are not yet available. It is therefore advisable to discuss the energy transfer for the two groups of stirrers in different sections.

3.1.1 Energy Transfer by Turbine and Paddle Stirrers

3.1.1.1 The Stirrer Model Concept

The successful analysis of energy transfer by stirrers to Newtonian and non-Newtonian fluids started with an observation made independently by the author and coworkers[5, 9)]

and by Nagata et al.[10]. According to this observation the liquid contained in the volume of rotation of a rotating stirrer behaves like a solid cylindrical body. The energy transfer in the cylinder-model takes place at the top and bottom surfaces of the cylinder and at the curved vertical surface. The energy transferred by the stirrer and across the surface of the model must be the same. As the fluid contained in the volume of rotation does not behave exactly like a solid body the diameter of the equivalent cylinder is less than that of the stirrer, i.e.:

(a) The height h_r of the cylinder and the stirrer are equal.

(b) The equivalent diameter d_{gl} of the cylinder is slightly smaller than that of the stirrer d_r.

In Fig. 18 a paddle stirrer and the equivalent cylinder are shown. The equivalent diameter d_{gl} is determined in such a way that the transferred energy is the same for both model and stirrer. The concept of the equivalent diameter was originally introduced by Thiele[11].

Using the cylinder-model it has been possible to solve the differential equations for the laminar flow field in mixing vessels by numerical methods and this has allowed the velocity field, the shear stress at the model surface and the transferred energy to be determined. The results so obtained are in excellent agreement with available experimental data in the laminar flow region. To extend the range of application into the turbulent flow region it has been necessary to introduce empirical correction terms leading to the general form of the energy transfer equation; i.e.

$$\mathrm{Ne} = f(\mathrm{Re}, d_r^*; d_s^*; d_w^*; h_r^*; h_s^*; h_w^*; l^*; b^*; H^*; s^*; k^*; c; i) \quad (5)$$

with the conditions: $\mathrm{Ga} \to \infty$ and where the dimensionless numbers are defined in Eqs. (7)-(17).

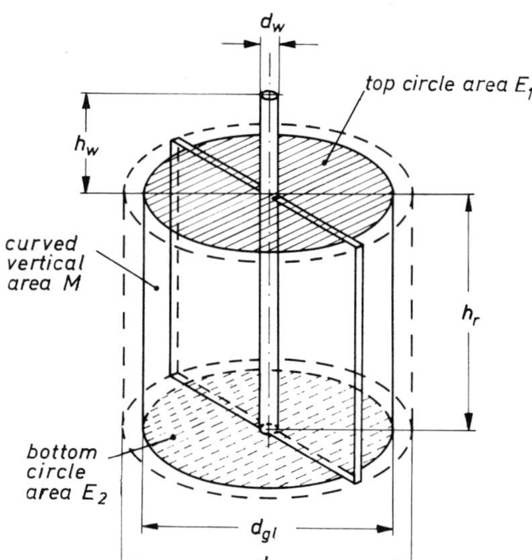

Fig. 18. Paddle stirrer and the equivalent stirrer model

The energy transfer equations were developed by Ihme[8)], who made use of the studies of Brauer[5)], Brauer and Thiele[12)], and Thiele[11)].

The general form of the equation for the upper limiting value of the Reynolds number Re_1 for vessels without baffles and Re_2 for vessels with baffles is as follows:

$$Re_i = f_i\,(Ga; Re; d_r^*; d_s^*; d_w^*; h_r^*; h_s^*; h_w^*; l^*; b^*; H^*; s^*; k^*; c; i) \tag{6}$$

The equations for Re_1 and Re_2 have been developed by Ihme[8)]. The definition of the dimensionless numbers and quantities are given below:

Ne see equation (1)	Newton number ,	
Re see equation (2)	Reynolds number ,	
Ga see equation (3)	Galilei number ,	
$d_r^* \equiv d_r/D$	Stirrer diameter ratio ,	(7)
$d_s^* \equiv d_s/D$	Plate diameter ratio ,	(8)
$d_w^* \equiv d_w/D$	Rotating shaft diameter ratio ,	(9)
$h_r^* \equiv h_r/D$	Stirrer height ratio ,	(10)
$h_s^* \equiv h_s/D$	Plate thickness ratio ,	(11)
$h_w^* \equiv h_w/D$	Rotating shaft height ratio ,	(12)
$H^* \equiv H/D$	Liquid height ratio ,	(13)
$l^* \equiv l/D$	Blade length ratio ,	(14)
$b^* \equiv b/D$	Blade thickness ratio ,	(15)
$s^* \equiv s/D$	Baffle length ratio ,	(16)
$k^* \equiv k/D$	Baffle thickness ratio ,	(17)
c	Number of blades and paddles ,	
i	Number of baffles .	

For dimensions of turbine and paddle stirrer see Figs. 2 and 3.

3.1.1.2 Energy Transfer Equation for Turbine and Paddle Stirrers in Vessels Without Baffles

The energy transfer equation for this case is:

$$Ne = \frac{C_1}{Re} + \left(\frac{1}{C_2 + C_3} - \frac{1}{C_2 + C_4\,Re^2}\right) + \frac{Re}{C_5 + C_6\,Re^{1.113}}\ . \tag{18}$$

The first term on the right hand side of Eq. (18) applies to the laminar region and is based on theoretical analysis. The second term applies to the transition region and the third term to the fully turbulent region. The last two terms are empirical terms based on extensive experimental data. The coefficients C_1 to C_6 are functions of geometric ratios given by:

$$C_1 = 4\pi^3 \left(\frac{d^*_{gl}}{d^*_r}\right)^3 \left[\frac{\{(0.6\, h_r^{*0.6})^5 + (1.3\, h_r^*)^5\}^{1/5}}{d^*_{gl}\,(1 - d^{*2}_{gl})}\right.$$

$$\left. + \frac{0.95\, h_w^{*1.4}}{d^*_w\,(1 - d^{*2}_w)} \left(\frac{d^*_w}{d^*_{gl}}\right) + \frac{6}{\pi^3}\, d_{gl}^{*1/3} \left\{2 - \left(\frac{d^*_w}{d^*_{gl}}\right)^4\right\}\right], \tag{19}$$

where d^*_{gl} is the equivalent diameter of the cylinder-model and is given by Eq. (20) for turbine stirrers and Eq. (21) for paddle stirrers:

$$d^*_{gl} = 1.09\, d^*_r\, b^{*0.015} \left[\left\{1 + \frac{1}{(167 + 0.00167\, c^4)^{0.5}\, d_r^{*1.375}}\right\}^{-4}\right.$$

$$\left. + \frac{1.417}{d_r^{*0.36}\, c^{0.44}}\right]^{-1/2}$$

$$\cdot \left[1 + \frac{1}{d_r^{*0.75}\left\{3.4324\left(\frac{1^*}{d^*_r}\right)^{0.24} + 1.4 \times 10^6 \left(\frac{1^*}{d^*_r}\right)^6\right\}^{5/3}}\right]^{-1/5}, \tag{20}$$

$$d^*_{gl} = 1.09\, b^{*0.015} \left[17.5\, h_r^{*0.004}\left(\frac{0.0708}{d_r^{*1.15}} - 1\right) + 17.5\right]^{-1}$$

$$\cdot \left[1 + \frac{1}{d_r^{*0.75}\left\{3.4\left(\frac{1^*}{d^*_r}\right)^{0.24} + 5 \times 10^5 \left(\frac{1^*}{d^*_r}\right)^6\right\}^{5/3}}\right]^{-1/5}. \tag{21}$$

The functions C_2 to C_6 apply to both types of stirrers:

$$C_2 = \left[\left\{\left(0.024\left[\frac{d^*_r}{h^*_r}\right]^{2.13}\right)^4 + (0.55\, d_r^{*2.35})^4\right\}\left\{\left(\frac{7.18}{c^{1.1}}\right)^4\right.\right.$$

$$\left.\left. + (0.001\, c^2)^4\right\}\right]^{1/4} \left[\left(0.138\,\frac{d^*_r}{1^*}\right)^{6.4} + 1\right]^{1/8}, \tag{22}$$

$$C_3 = \frac{0.0046\, d_r^{*\,0.735}}{c^{0.4}} \left[\frac{0.0009}{h^{*2}} + 0.36\right] \left[\left(0.069\,\frac{d_r^*}{l^*}\right)^{4.8} + 1\right]^{1/3}, \tag{23}$$

$$C_4 = 0.056\,\frac{h_r^{*\,0.1}}{c^{2.42}}, \tag{24}$$

$$C_5 = \left[\frac{1}{c^3}\left(\frac{2{,}6}{h_r^{*\,3}} + 1.5 \times 10^4\right)\right]^{1/3}, \tag{25}$$

$$C_6 = 3.35\, d_r^{*\,1.1} \left[\left\{\frac{2.77 \times 10^{-6}}{c^{2.56}\, h_r^{*\,4}} + \frac{8.64 \times 10^{-3}}{c^{0.96}\, h_r^{*\,0.52}}\right.\right.$$

$$\left.\cdot\left(1 + \frac{0.45}{c^{0.6}}\,\frac{[b^{*\,0.005} - 0.97]}{[0.004 + h_r^{*\,2.1}]}\right)^4\right\}^{1.25}$$

$$\left.+ \frac{0.0014}{c^{0.5}}\right]^{1/5} \left[\left(0.042\,\frac{d_r^{*\,0.5}}{l^*}\right)^4 + 1\right]^{1/4} [1 + 6 \times 10^{-6}\, c^4]^{0.1} \tag{26}$$

The range of application of Eqs. (18)–(26) as determined by experiment is as follows:

$$\begin{array}{lcccl}
0 & \leqslant & Re & \leqslant & 5 \times 10^5 \\
 & & Re & \leqslant & Re_1 \\
0.25 & \leqslant & d_r^* & \leqslant & 0.9 \\
0.025 & \leqslant & h_r^* & \leqslant & 0.9 \\
0.0175 & \leqslant & l^* & \leqslant & 0.45 \\
0.002 & \leqslant & b^* & \leqslant & 0.04 \\
2 & \leqslant & c & \leqslant & 12 \\
0.05 & \leqslant & h_w^* & \leqslant & 0.7 \\
0.2 & \leqslant & e^* & = & H^* - (h_r^* + h_w^*) \\
 & & H^* & = & 1
\end{array}$$

It is advisable to make use of a computer when the equations given are applied. For the most important parameters – h_r^*, d_r^*, and c – Fig. 19 shows the space of application. Two examples of comparisons between data calculated with Eq. (18) and measured values are given in Figs. 20 and 21. The maximum deviation between the calculated and measured values of Ne occurs in the transition region and may be as much as ±10%.

In Sect. 2.5 it was pointed out that the Reynolds number Re must not exceed the upper limiting value Re_1. On the basis of experimental data, an equation has been developed for Re_1:

$$\frac{Re_1^2}{Ga} = 0.24\, h_w^* \left[\frac{1}{d_r^{*6}} + 57\right]^{1/6} \left[\left(\frac{1}{h_r^{*6}} + \frac{6.87 \times 10^5}{h_r^{*1.5}}\right)^{1/3} \frac{h_w^{*2}}{c\, d_r^*\, Ga^{0.3}} + \frac{0.8}{c^{0.18}}\left(\frac{5.8 \times 10^{-4}}{h_r^{*5.6}} + 1\right)^{0.05}\right] \tag{27}$$

which is applicable over the ranges:

$10^4 \leqslant Ga \leqslant \infty$
$0.25 \leqslant d_r^* \leqslant 0.9$
$0.025 \leqslant h_r^* \leqslant 0.9$
$0 \leqslant h_w^* \leqslant 0.7$
$2 \leqslant c \leqslant 12$
$b = 0.00348$, $l^*/d_r^* = 0.25$, $H^* = 1$ } parameters of less influence

3.1.1.3 Energy Transfer Equation for Turbine and Paddle Stirrers in Vessels with Four Baffles and $s^* = 0.1$

The energy transfer equation for this case is given as follows:

$$Ne = \frac{C_1}{Re} + \left[\left\{\left(\frac{C_7}{Re}\right)^2 + \left(\frac{C_8}{Re^{0.1}}\right)^2\right\}^5 + C_9^{10}\right]^{-1/10} . \tag{28}$$

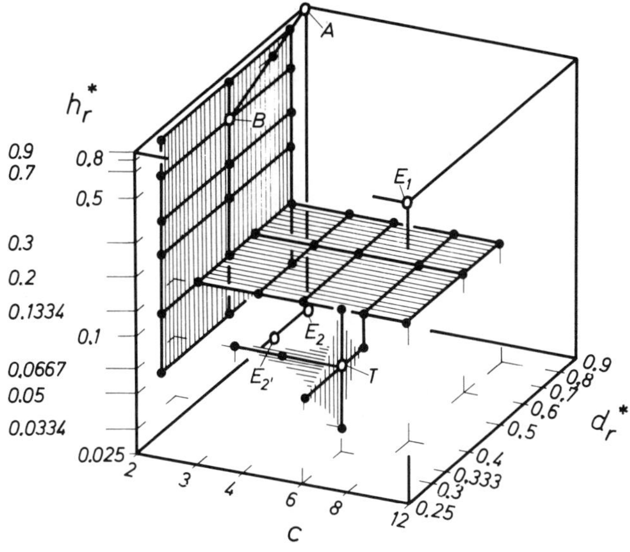

Fig. 19. Space of application for Eq. (18) with respect of the three most important parameters for turbine and paddle stirrers in vessels without baffles

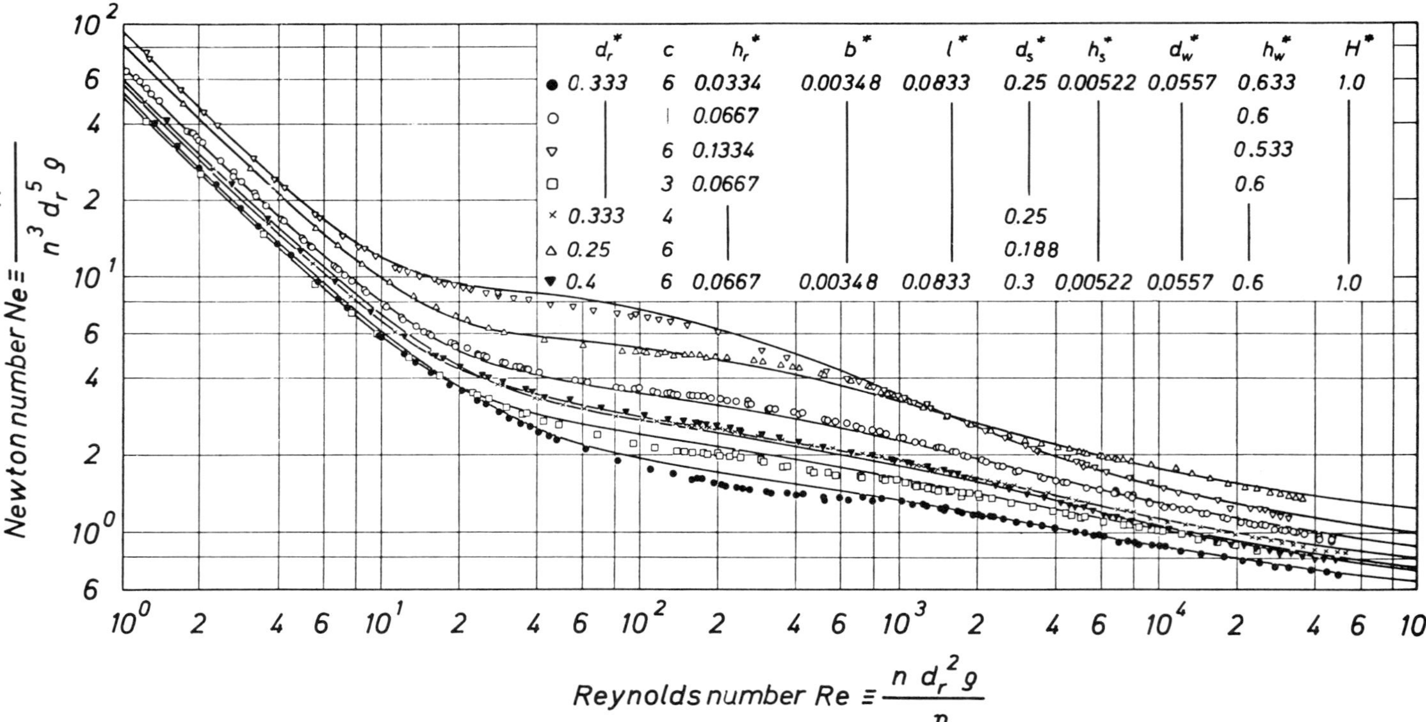

Fig. 20. Comparison between measured and calculated Ne-data for turbine stirrers in vessels without baffles. The curves have been calculated by means of Eq. (18)

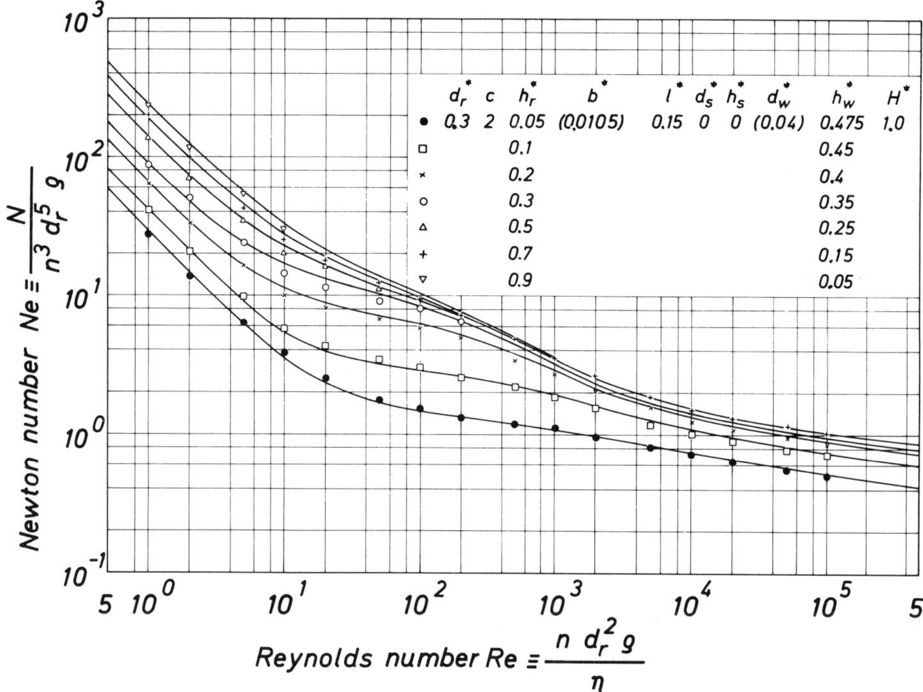

Fig. 21. Comparison between measured and calculated Ne-data for paddle stirrers in vessels without baffles. The curves have been calculated by means of Eq. (18)

The term C_1/Re describes the laminar flow region with C_1 given by Eq. (19), while the other two terms cover the transition and fully turbulent regions. The coefficients C_7 to C_9 are given by:

$$C_7 = \frac{0.195}{[(0.21\, h_r^{*\,0.35})^2 + (20.8\, h_r^{*\,2})^2]^{1/2}} \left[\left(\frac{4.7}{c}\right)^2 + 1\right]^{1/2} \cdot \frac{d_r^{*\,1.1}}{b^{*\,0.2}\,(0.19 + 7.6\, h_s^*)}\,, \tag{29}$$

$$C_8 = C_9\, \frac{0.27 + 0.45\left(\dfrac{h_s^*}{h_r^*}\right)^{0.25}}{\dfrac{0.13}{\left[1 + \left(4\,\dfrac{d_r^*}{c}\right)^5\right]^{1/5}} + 0.089\left(\dfrac{h_s^*}{h_r^*}\right)^{0.5} + 60\left(\dfrac{h_s^*}{h_r^*}\right)^5}\,, \tag{30}$$

$$C_9 = b^{*0.11}\, d_r^{*1.25} \left[\frac{7.1 \times 10^{-6}}{h_r^{*4.6}} + \left(\frac{0.22}{h_r^{*7}} + 80\right)^{0.2}\right]^{0.625}$$

$$\cdot \left(\frac{342}{c^4} + 0.098\right)^{0.25} \left[\frac{0.13}{\left\{1 + \left(5.5\,\frac{d_r^*}{c}\right)^5\right\}^{1/5}} + 0.089\left(\frac{h_s^*}{h_r^*}\right)^{0.5}\right.$$

$$\left. + 60\left(\frac{h_s^*}{h_r^*}\right)^5\right] \left[\left(\frac{0.22\, d_r^{*0.75}}{1^{*0.75}}\right)^{10} + 1\right]^{1/10} . \tag{31}$$

The range of application of Eq. (28) as determined by experiment is given as follows:

$0 \leqslant Re \leqslant \infty$
$Re \leqslant Re_2$
$0.25 \leqslant d_r^* \leqslant 0.6$
$0.03 \leqslant h_r^* \leqslant 0.6$
$0.05 \leqslant 1^* \leqslant 0.3$
$0.002 \leqslant b^* \leqslant 0.03$
$2 \leqslant c \leqslant 12$
$0.025 \leqslant h_s^*/d_r^* \leqslant 0.25$ and $h_s^* = 0$
$0.2 \leqslant h_w^* \leqslant 0.7$
$0.2 \leqslant e^* \leqslant H^* - (h_r^* + h_w^*)$
$H^* = 1$
$s^* = 0.1$
$k^* = 0.0174$ baffle characteristics
$i = 4$

The baffle thickness ratio $k^* \equiv k/D$ is a parameter of lesser importance and the equations may be used with accuracy for other values.

For application of Eq. (28) the use of a computer is advised. A space of application with respect to the three most important parameters is given in Fig. 22. Two examples of comparison between data calculated using Eq. (28) and measured values are given in Figs. 23 and 24.

In Sect. 2.5 it was pointed out that the Reynolds number must not exceed the upper limiting value Re_2 when the vessels are equipped with baffles. An equation has been developed for Re_2 using experimental data; namely:

$$Re_2 = Re_1 \left[\frac{0.376}{(h_r^{*0.9} + 7.5\, h_r^{*4})^{0.5}\, c^{0.22}} \left(\frac{4.8 \cdot 10^{-3}}{d_r^{*3}} + 1\right)\right]^{\frac{1}{2m}} , \tag{32}$$

where

$$m = 1 + \frac{2.8 \times 10^6}{Ga}$$

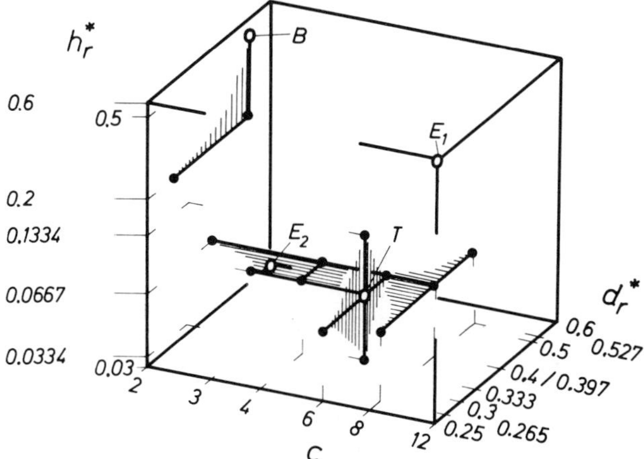

Fig. 22. Space of application for Eq. (28) with respect of the three most important parameters for turbine and paddle stirrers in vessels with baffles

and which is applicable over the ranges:

$$
\begin{aligned}
10^4 &\leqslant Ga \leqslant \infty \\
0.03 &\leqslant h_r^* \leqslant 0.6 \\
0.25 &\leqslant d_r^* \leqslant 0.6 \\
2 &\leqslant c \leqslant 12 \\
b^* &= 0.00348 \\
h_s^* &= 0.00522 \\
1^*/d_r^* &= 0.25 \\
H^* &= 1 \\
s^* &= 0.1 \\
k^* &= 0.0174 \\
i &= 4
\end{aligned}
$$

3.1.1.4 Discussion on the Influence of the Most Important Parameters for Energy Transfer

The most important parameters for energy transfer by turbine and paddle stirrers are the stirrer diameter ratio d_r^*, the stirrer height ratio h_r^* and the number of blades and paddles c. In Fig. 25 Ne is presented as a function of these three parameters for vessels without (o) and with (m) baffles. Figure 25 shows a remarkable influence of the stirrer height h_r^* on the Newton number when the vessel is equipped with baffles. As the desired fluid flow in the vessel is to be achieved with the lowest possible energy transfer it is clearly necessary to think very carefully before using baffled vessels.

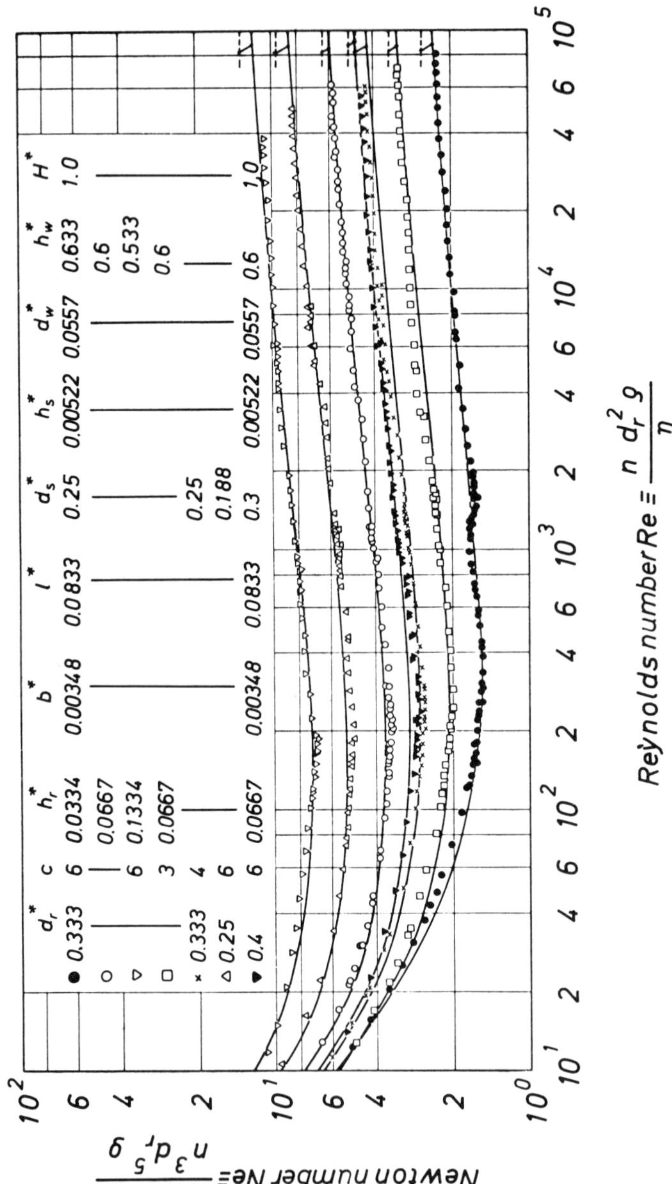

Fig. 23. Comparison between measured and calculated Ne-data for turbine stirrers in vessels with baffles. The curves have been calculated by means of Eq. (28)

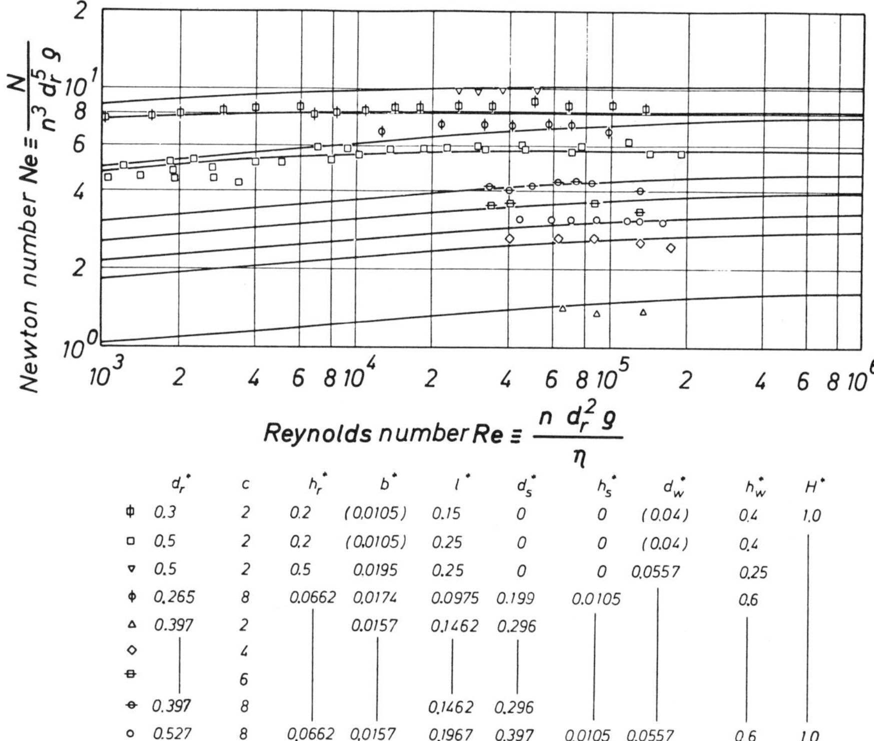

	d_r^*	c	h_r^*	b^*	l^*	d_s^*	h_s^*	d_w^*	h_w^*	H^*
◫	0.3	2	0.2	(0.0105)	0.15	0	0	(0.04)	0.4	1.0
□	0.5	2	0.2	(0.0105)	0.25	0	0	(0.04)	0.4	|
▽	0.5	2	0.5	0.0195	0.25	0	0	0.0557	0.25	|
Φ	0.265	8	0.0662	0.0174	0.0975	0.199	0.0105	|	0.6	|
△	0.397	2	|	0.0157	0.1462	0.296	|	|	|	|
◇	|	4	|	|	|	|	|	|	|	|
⊟	|	6	|	|	|	|	|	|	|	|
⊖	0.397	8	|	|	0.1462	0.296	|	|	|	|
○	0.527	8	0.0662	0.0157	0.1967	0.397	0.0105	0.0557	0.6	1.0

Fig. 24. Comparison between measured and calculated Ne-data for turbine and paddle stirrers in vessels with baffles. The curves have been calculated by means of Eq. (28)

3.1.2 Energy Transfer by Propeller and Disk Stirrers

The information available on energy transfer by propeller and disk stirrers suitable for gas dispersion is rather scarce. The results of experimental investigations carried out by Glaeser et al.[13)] on a vessel containing four baffles are presented in Figs. 26 and 27. The shape and dimensions of the stirrers are given in Figs. 4 and 5.

Figure 26 contains results for 1-, 2-, 3-, and 4-bladed propellers in the high turbulent flow region. The arrows indicate the upper limiting Reynolds number Re_2. In the high turbulent flow region the Newton number Ne is independent of Re and the results are correlated by the empirical equation:

$$Ne = 0.026\, c \quad . \tag{33}$$

For a 4-bladed propeller Ne = 0.104, which is about 2% of the Newton number for turbine stirrers.

Experimental data for disk stirrers are presented in Fig. 27. The number of the holes near the outer edge of the disk has been varied between z = 0 and z = 12, and in one case

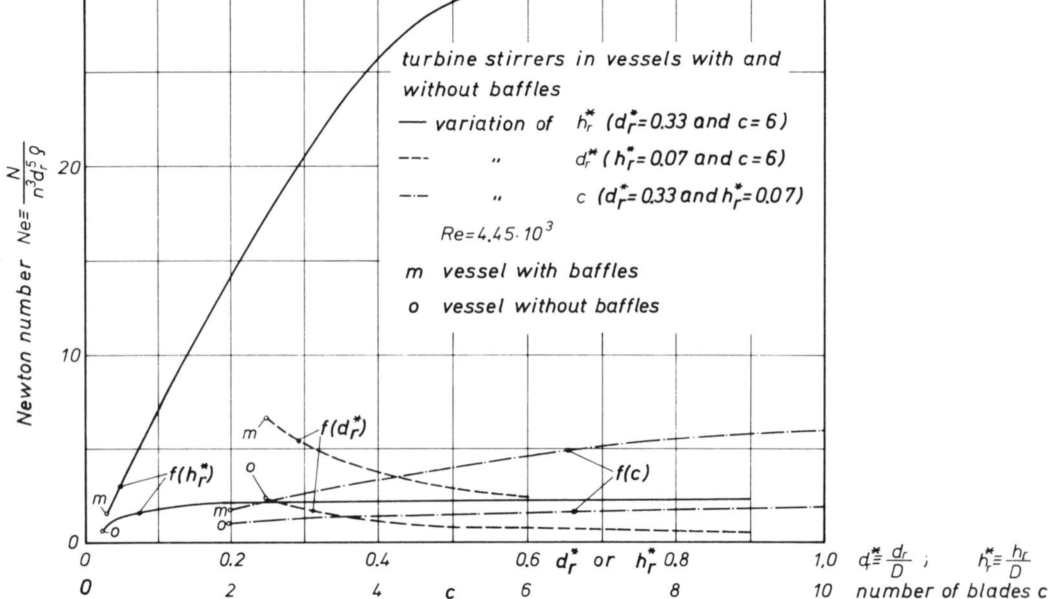

Fig. 25. Influence of the most important parameters on the Newton number Ne for turbine stirrers in vessels without and with baffles

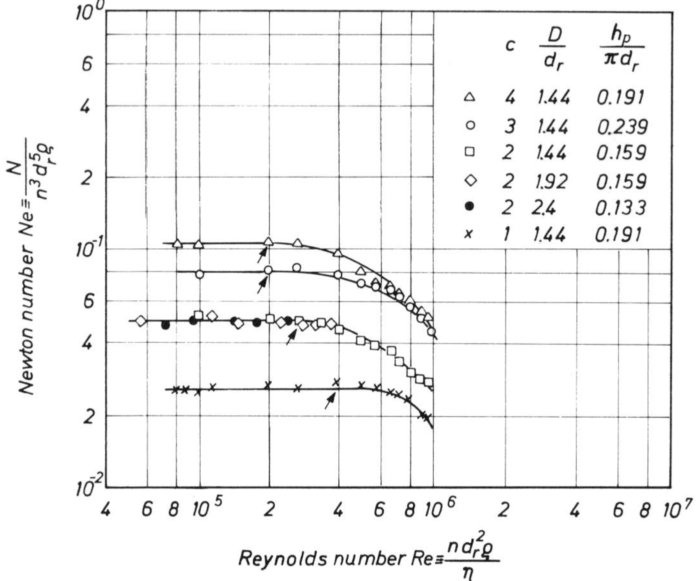

Fig. 26. Newton number Ne for 6 propeller stirrers in baffled vessels as a function of Reynolds number Re in the high turbulent region

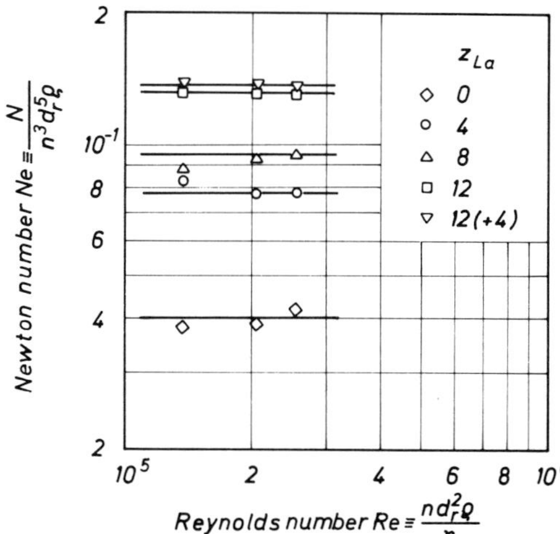

Fig. 27. Newton number Ne for 5 disk stirrers in baffled vessels as a function of Reynolds number Re in the high turbulent region

there were four more holes arranged on an inner circle. The diameter of the holes was 21 mm and the diameter ratio d_r/D was 0.5. Within the tested range the data are well correlated by the equation:

$$Ne = 0.04 + 0.0075\, z \; . \tag{34}$$

When z = 12 the Newton number is 0.13; a value less than 3% of that for turbine stirrers.

3.2 Energy Transfer in Pure Non-Newtonian Liquids

The viscosity of the majority of non-Newtonian liquids is relatively high. As a consequence the stirrer operations take place predominantly in the laminar flow region, so that the Reynolds number is in most cases below a value of roughly 10.

Currently information on energy transfer is restricted to power law fluids. For other fluids, especially when viscoelastic effects are present, it is necessary to rely on specific laboratory tests.

3.2.1 Energy Transfer Calculation Method after Metzner and Otto

Metzner and Otto[14] developed and tested a rough but simple method by means of which the energy transfer in non-Newtonian liquids can be calculated when information is available on energy transfer in Newtonian liquids for the same type of stirrer and vessel. The method involves the introduction of a characteristic viscosity η of the non-Newtonian liquid in the vessel under stirring conditions. Using this viscosity in the Reynolds number of the equations available for Newtonian fluids, these equations may be used to calculate Ne for the non-Newtonian fluid.

For purely viscous non-Newtonian fluids, i.e., the so-called power law fluids, the characteristic viscosity η is defined by the equation, that relates the shear stress τ with the velocity gradient:

$$\tau = k/\gamma/^{m-1} \quad , \tag{35}$$

where k is the consistency factor and m the fluid index.

For the mean velocity gradient or shear rate (γ) divided by the number of revolutions per unit time (n) Metzner and Otto found a constant value between 11.5–13 for turbine stirrers. Calderbank and Moo-Young[15] found that γ/n is a function of the stirrer diameter ratio d_r^*. In a still later paper Schilo[9] found that for paddle stirrers γ/n becomes independent of the diameter ratio d_r^* when d_r^* is smaller than 0.7 and is a function of the fluid index m. The results obtained by Schilo may be expressed by the empirical equation:

$$\frac{\gamma}{n} = 4.4 \, m^{-1.625} \quad . \tag{36}$$

The tested range of applicability is: $0.488 \leqslant m \leqslant 0.81$
$d_r^* \leqslant 0.7$.

It is suggested that Eq. (36) be used for turbine and paddle stirrers but that it should be used with caution in the case of propeller and disk stirrers. However the errors that may result from applying Eq. (36) are not very serious, since the energy transfer by propeller and disk stirrers is much smaller than that for turbine and disk stirrers.

3.2.2 Energy Transfer Calculation Method after Schilo

Schilo[9] has derived a theoretical equation for the Newton number based on the concept of the cylinder-model discussed in Sect. 3.1.1.1. However, comparison of the predicted data with experimental results has made it necessary to introduce an empirical correction. The equation presented by Schilo holds for the following conditions:

(a) power law fluids with $0 \leqslant m \leqslant 1$, i.e., pseudoplastic fluids
(b) paddle stirrers and
(c) laminar flow region .

The Shilo equation is as follows:

$$Ne = 0.7 \, \pi^2 \left(\frac{4\,\pi}{m}\right)^m \frac{(h_r/d_r)\, D^{*2}}{(D^{*2/m} - 0.75)^m} \, \frac{1}{Re_m^{0.9}} \quad . \tag{37}$$

Where the Reynolds number Re_m and D^* are defined by:

$$Re_m \equiv \frac{n^{2-m} \, d_r^2 \, \varrho}{k} \quad , \tag{38}$$

$$D^* \equiv 1/d_r^* = D/d_r \quad . \tag{39}$$

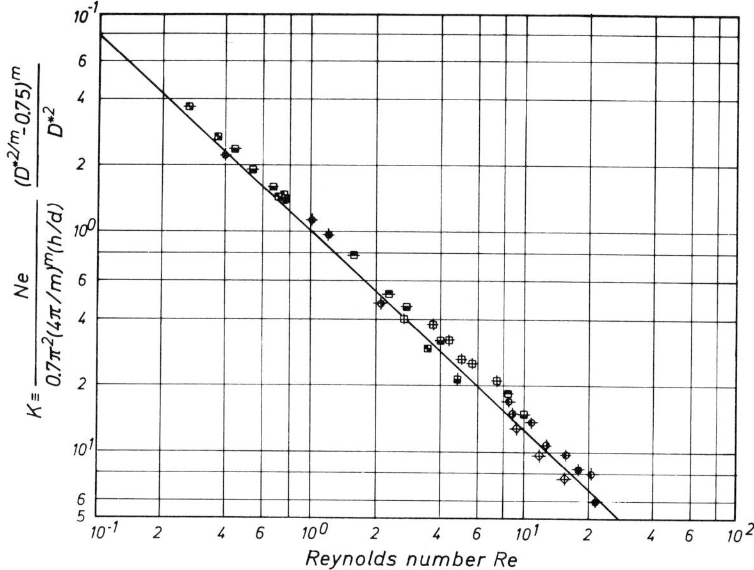

Fig. 28. Newton number Ne for paddle stirrers in power law fluids. Curve after Eq. (37)

In Eq. (38) k is the consistency factor of the shear stress/shear rate relationship defined by Eq. (35). Experimental data obtained by Schilo for paddle stirrers in power law fluids are presented in Fig. 28. The curve drawn through the data was calculated using Eq. (37).

4 Energy Transfer in Aerated Systems

Energy transfer in aerated systems is an extremely complicated physical process and there are no equations of general importance available for practical application. However the experimental information which has been gathered by Biesecker[3], van't Riet[4], and Glaeser et al.[13] assists understanding of some of the phenomena involved.

Figure 29 gives the Newton number for a 6-bladed turbine stirrer in a baffled vessel as a function of the gas Reynolds number Re_G. With increasing gas flow rate, while all other parameters are kept constant, the energy transfer from the stirrer to the two-phase fluid decreases. There are four distinct regions to be observed.

In region 1 at very low gas flow rates the gas collects beneath the stirrer resulting in a gas reservoir, from which single gas bubbles are torn away due to the liquid movement that is directed towards the periphery of the stirrer by the action of centrifugal forces. Observations reveal, that initially gas dispersion starts in the wake of one blade, while with increasing gas flow rate, gas dispersion takes place in the wake of two blades and so on.

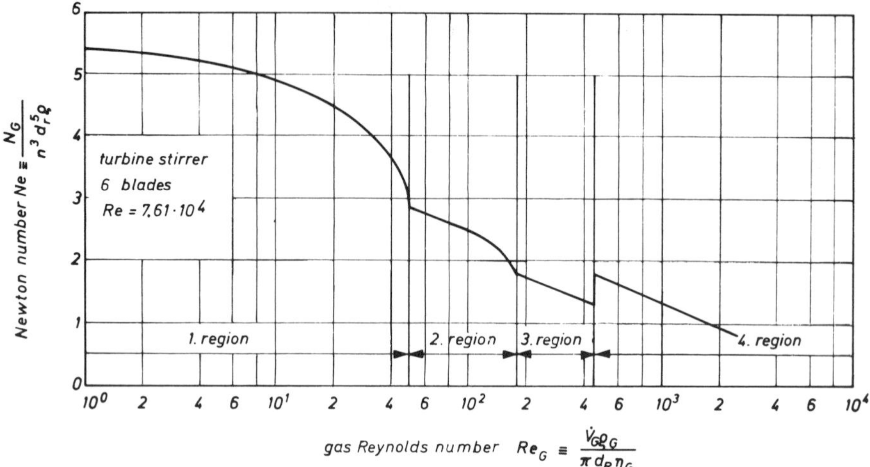

Fig. 29. Newton number Ne for a six-bladed turbine stirrer in a baffled vessel as a function of the gas Reynolds number Re_G

In region 2 increase of gas flow rate leads to the formation of vortex sheets. The procedure of sheet formation is the same as that for vortex thread formation in the region 1. At the end of region 2 vortex sheets have been built up behind all blades.

In region 3 there are stable vortex sheets behind all blades. This is the region with the highest gas dispersion efficiency and the energy transfer has been reduced to about 1/3 of that when $V_G = 0$.

In region 4 the increased gas flow rate leads to flooding of the stirrer. This is characterized by the formation of very big bubbles and the fact that gas dispersion process becomes ineffective.

Similar observations have been made when other types of stirrers are used as dispersion elements. Some further examples are given in Figs. 30 and 31 for propeller and

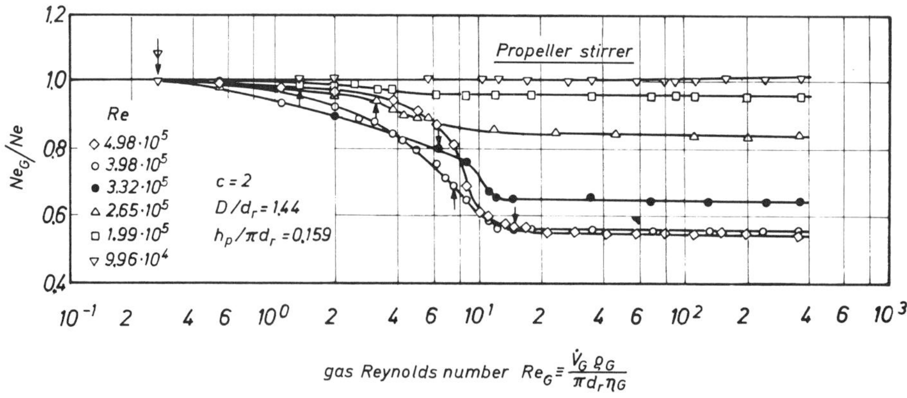

Fig. 30. Ratio of Newton numbers for propeller stirrers in aerated systems

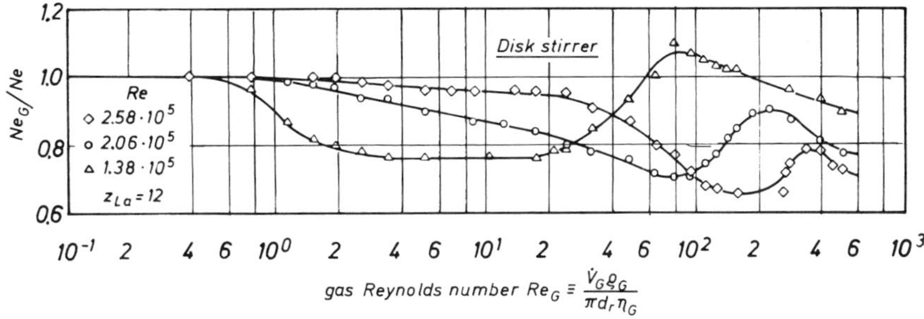

Fig. 31. Ratio of Newton numbers for disk stirrers in aerated systems

disk stirrers. In these figures the ratio of Newton numbers Ne_G/Ne is given as a function of the gas Reynolds number Re_G. Ne_G is the Newton number for the aerated and Ne for the unaerated system.

On the basis of available evidence it seems that energy transfer is reduced by the presence of a gas by about
40% for turbine stirrers,
40% for paddle stirrers,
30% for propeller stirrers,
10% for disk stirrers.

It seems reasonable to suggest that these values are also applicable to non-Newtonian fluids[16].

5 Symbols

b	thickness of blade and paddle
d_k	diameter of hole centre line of disk stirrer
d_L	hole diameter of disk stirrer
d_r	stirrer diameter
d_s	plate diameter of turbine stirrer
D	vessel diameter
e	distance between stirrer and bottom of vessel
g	gravitational acceleration
h_p	pitch of propeller
h_r	height of stirrer
h_s	plate stickness of turbine stirrer
h_w	height of rotating axis in liquid
H	liquid height of vessel
k	consistency factor
l	length of blade and paddle
m	fluid index
n	number of revolutions
N	energy transferred by the stirrer to the fluid

s		radial length of baffle
w		local velocity
γ		shear rate
η		fluid viscosity
ϱ		fluid density
τ		shear stress
b^*	$\equiv b/D$	blade thickness ratio
c		number of blades and paddles
d_r^*	$\equiv d_r/D$	stirrer diameter ratio
D^*	$\equiv 1/d_r^*$	vessel diameter ratio
d_s^*	$\equiv d_s/D$	plate diameter ratio
d_w^*	$\equiv d_w/D$	rotating shaft diameter ratio
Ga	$\equiv \dfrac{d_r^3\, g\, \varrho^2}{\eta^2}$	Galileo number
h_r^*	$\equiv h_r/D$	stirrer height ratio
h_s^*	$\equiv h_s/D$	plate thickness ratio
h_w^*	$\equiv h_w/D$	rotating shaft height ratio
H^*	$\equiv H/D$	liquid height ratio
i		number of baffles
k^*	$\equiv k/D$	baffle thickness ratio
l^*	$\equiv l/D$	blade length ratio
Ne	$\equiv \dfrac{N}{d_r^5\, n^3\, \varrho}$	Newton number
Re	$\equiv \dfrac{n\, d_r^2\, \varrho}{\eta}$	Reynolds number
Re_m	$\equiv \dfrac{n^{2-m} d_r^2\, \varrho}{k}$	Reynolds number for power law fluids
s^*	$\equiv s/D$	baffle length ratio
z		number of holes in disk stirrers

6 References

1. Taguchi, H.: The nature of fermentation fluids. In: Adv. Biochemical Eng., Vol. 1., p. 1. Berlin, Heidelberg, New York: Springer 1971
2. Brauer, H.: Grundlagen der Einphasen- und Mehrphasenströmungen. Aarau, Frankfurt: Sauerländer 1971
3. Biesecker, B.O.: Begasen von Flüssigkeiten mit Rührern. VDI-Forschungsheft 554. Düsseldorf: VDI-Verlag GmbH 1972
4. van't Riet, K.: Turbine agitator hydrodynamics and dispersion performance. Dissertation, T. H. Delft 1975
5. Brauer, H.: Chem.-Ing.-Techn. *39*, 209 (1967)
6. Brauer, H., Mewes, D.: Chem.-Ing.-Techn. *45*, 461 (1973)
7. Brauer, H., Schmidt-Traub, H.: Chem.-Ing.-Techn. *44*, 1237 (1972)
8. Ihme, F. Fortschr.-Ber. BDI-Z. Reihe 7, Nr. 44
9. Schilo, D.: Leistungsbedarf beim Rühren nicht-Newtonscher Flüssigkeiten. Dissertation D 83, Techn. Univ. Berlin 1968

10. Nagata, S., Yamamoto, K., Hashimoto, K., Naruse, Y.: Mem. Fac. Eng. Kyoto University *21*, 260 (1959)
11. Thiele, H.: Strömung und Leistungsbedarf beim Rühren Newtonscher Flüssigkeiten mit Anker-, Blatt- und Turbinenrührern im laminaren Bereich. Dissertation, Technische Universität, Berlin 1972
12. Brauer, H., Thiele, H.: Grundlagen für die Dimensionierung von Rührern und Rührgefäßen. Dechema Monographien, p. 69. Weinheim/Bergstr.: Verlag Chemie GmbH 1971
13. Glaeser, H., Biesecker, B., Brauer, H.: Verfahrenstechnik *7*, 31 (1973)
14. Metzner, A.B., Otto, R.E.: A.I.Ch.E. Journal *3*, 3 (1957)
15. Calderbank, P.H., Moo-Young, M.B.: Trans. Instn. Chem. Engrs. *39*, 337 (1961)
16. Höcker, H., Langer, G.: Rheol. Acta *16*, 400 (1977)

Loop Reactors

Heinz Blenke
Institut für Chemische Verfahrenstechnik
Universität Stuttgart, D-7000 Stuttgart 1, West Germany

1 Demands on Bioreactors for Fluid Biosystems 122
2 Principles and Types of Loop Reactors (*LR*) 124
3 Fluid Dynamics of Loop Reactors (*LR*) 128
3.1 Optimal Fluid Dynamic Design of *LR* for Aqueous (Quasi-)Homogeneous Liquid-Systems (*L*-Systems) 128
3.2 Flow Behavior of *PLR* and *JLR* with Highly Viscous (Quasi-)Homogeneous *L*-Systems 137
3.2.1 Experimental Fluids 137
3.2.2 Experimental Equipment 138
3.2.3 Theoretical Relations for Newtonian Fluids 138
3.2.4 Some Important Results for Newtonian Fluids 142
3.2.5 Power Demand and Circulation in *PLR* for Non-Newtonian (Pseudoplastic) Fluids . 146
3.3 Flow Behavior of Heterogeneous Solid-Liquid-Systems (*S–L*-Systems or Suspensions) in *JLR* 148
3.4 Flow Behavior of Heterogeneous Gas-Liquid-Systems (*G–L*-Systems) in *JLR* and *ALR* 151
3.4.1 Air-Water-System with *JLR* and *ALR* at $w_G \lesssim 10$ cm s^{-1} 151
3.4.2 Sulfite-System with *ALR* at $w_G \lesssim 60$ cm s^{-1} 155
4 Gas Hold up in *LR* 159
4.1 Introduction 159
4.2 Gas Hold up in *JLR* 159
4.3 Gas Hold up in *ALR* 162
5 Mixing and Residence Time Behavior of *LR* 162
5.1 Basic Principles 162
5.2 Degree of Mixing and Mixing Time 167
5.3 Residence-Time Behavior 172
6 Gas-Liquid-Interfacial Area (*G–L*-Interface) 176
6.1 Oxygen Transfer in Aerobic Bioreactions 176
6.2 *G–L*-Interface in *JLR* and *ALR* 182
6.3 *JLR* with Reversed Flow Direction 189
7 Heat Transfer, Limiting Capacity, and Stability Behavior of *LR* 194
8 Economic Optimal Size of *LR* 198
9 Examples of Actual Research in the Development of *LR* 201
9.1 Optimal O_2-Conversion 201
9.2 Optimal Design, Operation, and Combinations of *LR* 203
10 Summary and Outlook 205
11 Nomenclature 207
12 References 213

This report on *advances* in a new technological development mostly confines to contributions of the author and his collaborators. One should not consider this as one-sided overvaluation but rather understand it as a restriction to originative cognition and opinion within a many-sided development and a variety of ideas.

Loop reactors are characterized by a definitely directed circulation flow, which can be driven in fluid or fluidized systems by propeller or jet drive and mainly in *G–L*-systems furthermore by "air-lift" drive. Considering biotechnology, they are especially appropriate for fluid systems requiring high dispersion effects, defined flow conditions throughout the whole reaction space and high specific cooling capacity. Their simple construction and operation and their well directed flow result in relatively low investment and operational costs. In consequence they can be considered as appropriate bioreactors for multiphase even highly viscous biosystems. But their optimal design and operation for large scale production plants still requires considerable research and development of chemical engineers in close cooperation with microbiologists and biochemists. This paper may give some contributions from an engineers point of view.

1 Demands on Bireactors for Fluid Biosystems

In this treatise the loop reactor (*LR*) will be discussed fundamentally with respect to its present state of development, considering specially its suitability as a bioreactor. Thus initially some of the most important demands for the technical realization of bioreactions in fluid systems shall be briefly presented. We consider here as an example the production of single cell protein (*SCP*), because we are just taking part in the development of appropriate *LR*-types for that[1, 2].

Such systems consist of an aqueous "culture broth" in which the single cells (yeast or bacteria) shall increase as rapidly as possible at a cell concentration $x \approx 20$ kg m^{-3} [a] broth ($\approx 2\%$) and at temperatures $T_R \approx 35$–40 °C, thus forming high-value biomass. This may occur, for example, at a *specific growth rate* $\mu \approx 0.2$ kg kg^{-1} h^{-1}, i.e., a specific *productivity of cell mass* $p = x\,\mu \approx 4$ kg m^{-3} h^{-1}.

Thus, as shown schematically in Fig. 1, substrate (e.g., methanol) is introduced at a specific mass flow rate $\dot{m}_{Sub} \approx 8$ kg m^{-3} h^{-1} where the methanol concentration should be $c_{Sub} \approx 1‰$ throughout the whole culture broth and nowhere exceed the mortality level $\hat{c}_{Sub} \approx 1\%$. This puts high demands on intense mixing within the whole reaction system.

In addition it is very important for such multi-phase biosystems, to avoid foaming as well as flotation of light and sedimentation of heavy compounds. In *LR* this can very effectively be achieved without chemical additives or special mechanical devices just by the suction or fluidization effects of the suitably directed circulation flow itself.

The high specific O_2-requirement $\dot{m}_{O_2} \approx 8$ kg m^{-3} h^{-1} of this aerobic bioreaction is of special importance. It demands a large specific *G–L*-interface for the O_2-transfer from the air (*G*) into the broth (*L*), i.e., for submerse processes an intensive dispersion of the air within the broth. Furthermore, using a defined air input, as large an average driving concentration difference as possible between gas and liquid phase should be

---

[a] The unity of volume is always referred to the liquid phase (e.g., culture broth), unless otherwise indicated.

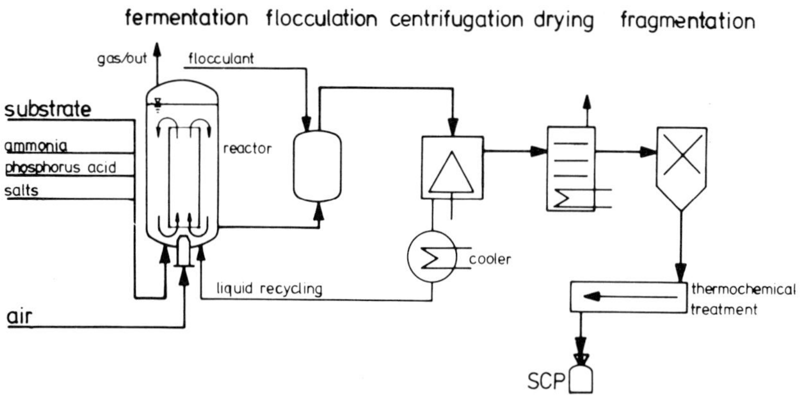

Fig. 1. Simplified flow sheet of an *SCP*-plant

aimed at. However the O_2-concentration in the broth must nowhere fall below $\check{c}_{O_2} \triangleq \check{c}_L \approx 1\ \mathrm{g\ m^{-3}}$ ($\triangleq$ 1 ppm). That means that by intensive dispersion of air and its defined transport throughout the reaction space a sufficient O_2-supply must always be assured everywhere.

The bioreaction is exothermic and approximately a specific thermal power $\dot{q} \approx 40\ \mathrm{kW\ m^{-3}}$ must be removed from the broth. This requires high heat transfer coefficients and large specific cooling areas in the reactor in order to operate, if possible, without external cooling despite the very small temperature difference $\Delta T_m \approx 8$ K available.

The necessary mixing, dispersing, and cooling effects should be achieved at the lowest possible power input (energy costs). This requires effective dispersion equipment and favorable flow guide to avoid unnecessary energy dissipation by useless vorticity.

Industrial production furthermore requires safe (sterility, environment) and economic (investment, energy, maintenance costs) large scale reactors ($V_R \approx 500\ \mathrm{m^3}$) with simple construction and operation.

With respect to favorable design, operation, and scale up all processes within the reactor determining its behavior should be calculable as far as possible, so that less experimental work is required in large bio-pilot plants to transfer results from small model apparatuses to large scale industrial production plants and from simple (if possible, non-biological) model systems to real bioproduction systems.

Finally, bioreactors should easily be adaptable to changes in operational parameters for temporarily differing operating conditions in batch processes, as well as for product or production changes in flow processes.

It arises already from this short discussion that certain demands on such bioreactors contradict each other. Thus, for example, a high mixing effect is best obtained with an *ideal stirred tank reactor* type, whereas the demand of high driving concentration difference for the O_2-transfer is best fulfilled with an *ideal tube reactor* type. We shall see that *LR* offer favorable combinations of both.

2 Principles and Types of Loop Reactors (LR)

LR are (bio-)chemical reactors, in which as shown in Fig. 2, at least one definitely directed circulation flow $\dot{M}_2$ is produced, which can be superimposed – e.g., in continuous operation – by a through flow $\dot{M}_1$ resulting in the total flow $\dot{M}_3 = \dot{M}_1 + \dot{M}_2$. In this way the flow pattern of a loop develops as presented especially clearly in Fig. 2b. Hence the name.

Figure 2a shows the principle of an *LR* with "internal circulation" around a draft tube, which is positioned concentrically in a slim tower reactor. It directs the coaxial circulation flow throughout the whole reactor volume V_R. The principle of an *LR* with "external circulation" is shown in Fig. 2b.

Loop apparatuses were already patented at the end of the last century to mix fluid or fluidized contents of slim tower type containers, for which the well-known principle of stirred tanks is unsuitable.

LR can be classified according to the mode of flow drive into the following types[3]:

- *Air-lift loop reactors (ALR)* with hydrostatic flow drive, caused by different densities – especially different gas hold up – of fluids in communicating spaces. Figure 3 shows such a design with internal circulation according to Fig. 2a[4]. In Fig. 4 an *ALR* with external circulation as in Fig. 2b is presented, known as the *ICI pressure cycle-fermenter*[5, 6].
  All types of *ALR* are distinguished by their very simple construction and operation. They are especially suitable for high gas flow rates and – in biotechnology – for sensitive organisms.
- *Propeller loop reactors (PLR)* with hydromechanic propeller (or pump) flow drive. This mode of impulsion is illustrated in Fig. 5, demonstrating at the same time, that the draft tube can also be used for heat transfer, which has far reaching consequences on the maximum size of reactors for high demands on heat transfer[7] (see Sect. 7). The *PLR* necessitates a more complicated construction (propeller drive, vibrations, foundations ...), causes operating problems related to the sealing of the shaft (leaks of poisonous or inflammable substances; sterility) and in biosystems possibly damage of sensitive organisms in the high shear fields at the propeller. In principle, a *PLR* is a slim stirred tank reactor (*STR*) with a longitudinal driving propeller-agitator. It is especially suitable for highly viscous fluids and small-to-medium size units.

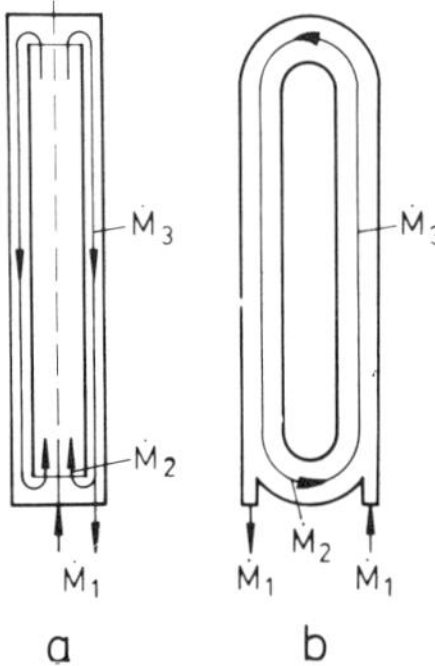

Fig. 2. Principles of loop reactors (*LR*)

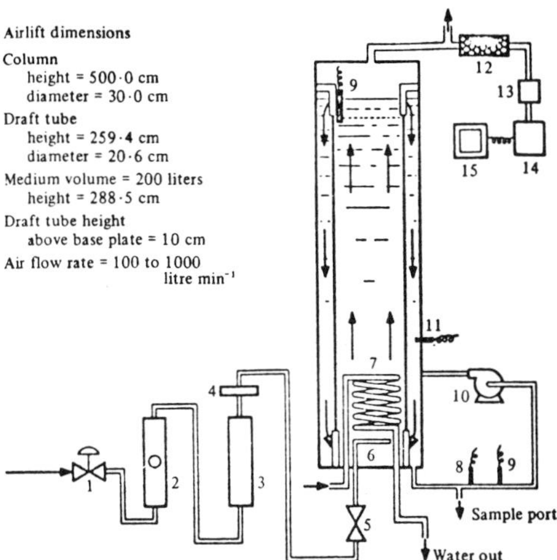

Air-lift fermenter (arrows indicate direction of liquid flow) (Wang and Humphrey, 1969).

1 Air-pressure regulator
2 Rotameter
3 Glass-wool air filter
4 Millipore air filter
5 Check valve
6 Air sparger
7 Heating and cooling coil
8 pH electrode
9 Oxygen probe
10 Centrifugal pump
11 Thermistor probe
12 Silica-gel bed
13 Diaphragm pump
14 Paramagnetic oxygen analyser
15 Recorder

(5) An air line terminating in a nozzle.

Fig. 3. Schematic figure of an air-lift loop reactor (*ALR*)

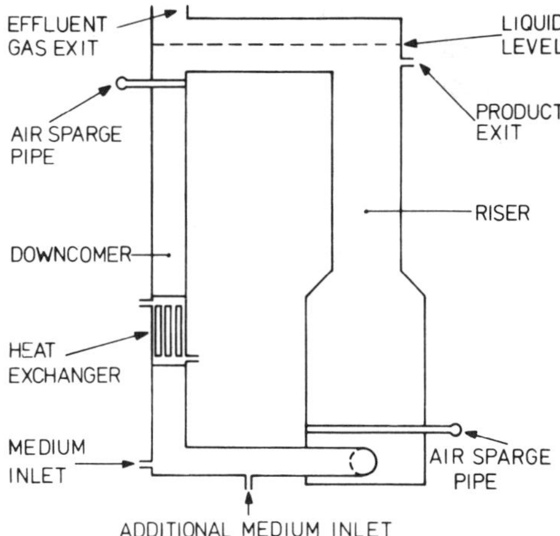

Fig. 4. *ICI* pressure cycle fermenter

- *Jet loop reactors (JLR)* with hydrodynamic jet flow drive. As shown in Fig. 6, for example, the liquid jet of mass flow $\dot{M}_1$ is injected with high velocity $w_1 \gtrsim 20$ m s^{-1} through at least one liquid nozzle, here at the bottom of the reactor. $\dot{M}_1$ drives the circulation flow $\dot{M}_2$ by momentum transfer and flows continuously out of the system as superimposed through flow. In batch operation it is completely recycled to the

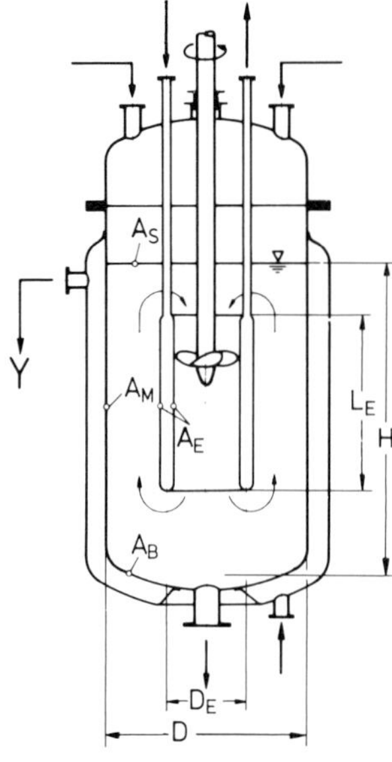

Fig. 5. Propeller loop reactor (*PLR*)

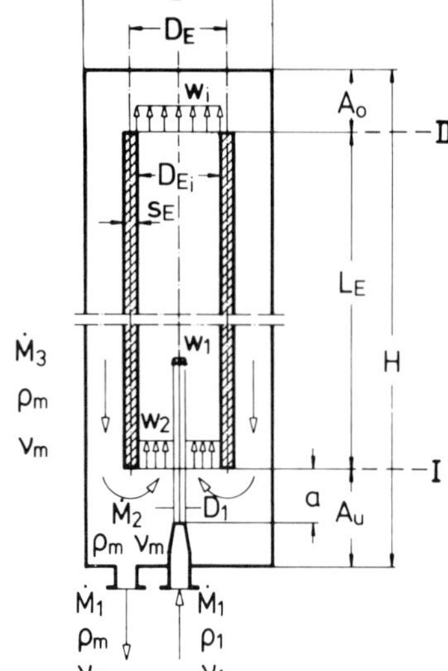

D : internal diameter of loop reactor
H : filling hight of reaction system
D_E: mean diameter of draft tube
D_{Ei}: internal diameter of draft tube
L_E: length of draft tube
A_u: distance of draft tube from bottom
A_o: distance of draft tube from surface
a : distance of draft tube from liquid nozzle
D_1: diameter of liquid nozzle and liquid jet
$\dot{M}_1$: liquid mass trough flow
$\dot{M}_2$: liquid circulation mass flow
$\dot{M}_3$: total mass flow of liquid

Fig. 6. Jet loop reactor (*JLR*) here completely filled with a liquid system (*L*-system)

liquid nozzle by an external closed pump cycle. In continuous operation the required driving through flow $\dot{M}_1$ may be greater than the throughput resulting from reaction mass balance. Then the difference between both flows has to be recycled.
In Fig. 7 gas (index G) is introduced at the same time through a ring nozzle arranged concentric to the liquid jet (index L). In such heterogeneous G–L-systems the liquid

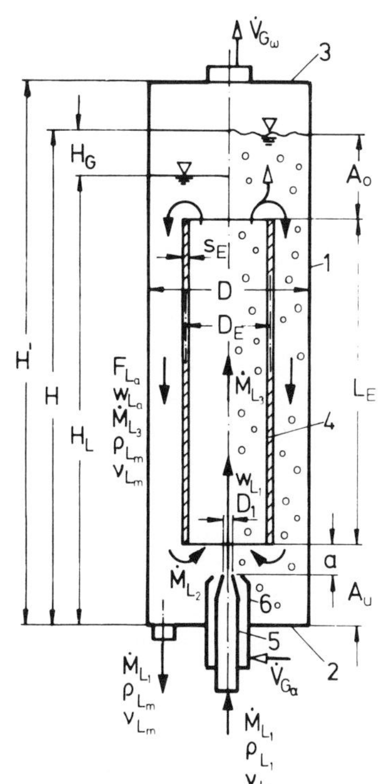

Fig. 7. Jet loop reactor (*JLR*) with a gas-liquid system (right), and without aeration (left)

jet causes not only the circulation drive, but also very efficient primary dispersion of the inlet gas, or of injected insoluble liquid phase, as for instance paraffin as substrate in the *SCP* culture broth. In addition, in *PLR* and *JLR* there is a very effective *re*dispersion of coalesced gas bubbles, liquid drops and solid agglomerates (e.g., cells) by the guided recirculation through the intense shear fields of the propeller or the liquid jet. In *G–L*-systems air-lift and jet drive superimpose each other. The more uniformly the gas is distributed in the whole space, the more jet drive predominates. On account of its simple construction and operation (no moving parts; mild hydrodynamic power input) and its defined mixing and intense dispersing effects with relatively low power requirements, the *JLR* is particularly suitable for heterogeneous *G–L*-systems with high demands for mass and heat transfer, as well as for defined mixing throughout the whole reaction space.

3 Fluid Dynamics of Loop Reactors (LR)

3.1 Optimal Fluid Dynamic Design of LR for Aqueous (Quasi-)Homogeneous L-Systems

Although in chemical engineering and biotechnology *LR* are especially suitable for heterogeneous *G–L*-systems, first of all the basic principles and the most important fluid dynamic parameters and relations shall be demonstrated briefly using the fundamental homogeneous liquid system of drinking water. In principle these considerations are approximately valid too for multi-phase systems with such a fine dispersion and uniform distribution of all components that the systems can be treated as *quasi-homogeneous*. In any case some methods and results from investigations of homogeneous systems can be transferred to heterogeneous systems too and others allow instructive comparisons between both of them.

Especially important parameters for optimal fluid dynamic design are (see Fig. 6) the following, where all dimensions are related to the reactor diameter D and all mass flows and velocities to the through flow $\dot{M}_1$ [3, 8]:

- height-to-diameter ratio of the reactor *(grade of slenderness)*

$$s \equiv \frac{H}{D} \tag{1}$$

- effective reactor volume

$$V_R = \frac{\pi}{4} D^2 H = \frac{\pi}{4} D^3 s \tag{2}$$

- reaction mass

$$M_R = \rho_m V_R = \rho_m \frac{\pi}{4} D^3 s \tag{3}$$

- circulation number

$$n_U \equiv \frac{\dot{M}_3}{\dot{M}_1} = \frac{\dot{M}_1 + \dot{M}_2}{\dot{M}_1} = 1 + \frac{\dot{M}_2}{\dot{M}_1} \tag{4}$$

- mean circulation velocity defined independent of D_E [3)]

$$w_m \equiv \frac{8 \dot{V}_3}{\pi D^2} = \frac{8 \dot{M}_3}{\rho_m \pi D^2} = \frac{8 \dot{M}_1}{\rho_m \pi D^2} n_U \sim n_U \tag{5}$$

- circulation rate from (3) and (5) with (1)

$$r_U \equiv \frac{\dot{M}_3}{M_R} = \frac{w_m}{2H} \equiv \bar{t}_U^{-1} \tag{6}$$

- mean circulation time

$$\bar{t}_U \equiv \frac{M_R}{\dot{M}_3} = \frac{2H}{w_m} = r_U^{-1} \tag{7}$$

- mean residence time of the continuous through flow $\dot{M}_1$ in V_R for liquid systems

$$\bar{t} \equiv \frac{M_R}{\dot{M}_1} = \frac{n_U}{r_U} = n_U \; \bar{t}_U \tag{8}$$

- liquid velocity (for JLR) at nozzle outlet

$$w_1 = \frac{4\,\dot{V}_1}{\pi\,D_1^2} = \frac{4\,\dot{M}_1}{\rho_1\,\pi\,D_1^2} \sim D_1^{-2} \tag{9}$$

- nozzle Reynolds number with (9)

$$Re_1 \equiv \frac{w_1\,D_1}{\nu_1} = \frac{4\,\dot{M}_1}{\nu_1\,\rho_1\,\pi\,D_1} \sim D_1^{-1} \tag{10}$$

- mean circulation Reynolds number with (4) and (5)

$$Re_m \equiv \frac{w_m\,D}{\nu_m} = \frac{8\,\dot{V}_3}{\nu_m\,\pi\,D} = \frac{8\,\dot{M}_1\,n_U}{\nu_m\,\rho_m\,\pi\,D} \sim \frac{n_U}{D} \sim w_m\,D \tag{11}$$

- circulation Reynolds ratio with (10) and (11)

$$\frac{Re_m}{Re_1} \equiv 2\,\frac{\nu_1\,\rho_1}{\nu_m\,\rho_m}\,\frac{D_1}{D}\,n_U \sim \frac{D_1}{D}\,n_U \tag{12}$$

or for constant geometrical parameters

$$\left(\frac{Re_m}{Re_1}\right)_{D_1/D} \sim n_U\,. \tag{13}$$

It is obvious that the contents of LR are the more intensely mixed, the larger the mean circulation velocity w_m, or – for certain values of D and $\dot{M}_1$ – the higher the circulation Reynolds number Re_m (11) resp. the circulation number n_U (5) is. The aim of this fluid dynamic optimization, which is demonstrated here for a *JLR*, is thus to achieve the fastest possible circulation flow with a given liquid jet power input P_L (30).

Decisively important for the intensity of the circulation flow which can be induced by P_L is the *resistance number* ζ_U (resp. the Euler number *Eu*) of the circulation flow, which can be expressed as the ratio of the pressure drop Δp_U (due to one circulation) to the dynamic pressure of the mean circulation velocity w_m:

$$\zeta_U \equiv \frac{2\,\Delta p_U}{\rho_m\, w_m^2} \equiv 2\,Eu_U\,. \tag{14}$$

The circulation flow, achieved by the momentum transfer from the injected liquid jet is described by the theorem of momentum. It states for the space considered inside the draft tube between balance areas I (inlet) and II (outlet) in Fig. 6: The difference between the inlet and outlet momentum flows $\dot{I}_I - \dot{I}_{II}$ is equal to the force transferred, and thus at steady state equal to the sum of all opposing resistance forces F_{I-II}

$$\dot{I}_I - \dot{I}_{II} = -\,F_{I-II} \tag{15}$$

or

$$\dot{M}_1\, w_1 + \dot{M}_2\, w_2 - \dot{M}_3\, w_i = \frac{\pi}{4}\, D_{E_i}^2 \Delta p_U = \frac{\pi}{4}\, D_{E_i}^2\, \zeta_U \frac{\rho_m}{2}\, w_m^2\,. \tag{16}$$

Using the variables mentioned above and with justifiable simplifications, one obtains from this in good approximation (ζ_U up to about 5% too large) the simple equation for the circulation resistance number ζ_U, which is to be minimized[8)]:

$$\zeta_U \approx 0{,}5\,\frac{\rho_m}{\rho_1}\left(\frac{D_1}{D}\,n_U\right)^{-2}\left(\frac{D_E}{D}\right)^{-2}. \tag{17}$$

Substituting $\frac{D_1}{D}\,n_u$ from (17) in (12), it follows

$$\frac{Re_m}{Re_1} \approx \frac{\nu_1}{\nu_m}\left(2\,\frac{\rho_1}{\rho_m}\right)^{\frac{1}{2}} \zeta_U^{-\frac{1}{2}}\left(\frac{D_E}{D}\right)^{-1}. \tag{18}$$

So much for the theory. Now one should only be able to calculate ζ_U in dependence of all material, geometrical, and fluid dynamic parameters, then n_U (17) and Re_m (12, 18) could be calculated and optimized in a theoretical way alone. According to the laws of fluid dynamics, ζ_U can be expressed as the sum of the consecutive partial resistances ζ_i and ζ_a for the internal and external *tubular flow* and ζ_o and ζ_u for the flow around the upper and lower edges of the draft tube, in this way defined as

$$\zeta_U^* \equiv \zeta_i + \zeta_a + \zeta_o + \zeta_u\,. \tag{19}$$

ζ_i and ζ_a can be calculated from well-known fluid dynamic laws. However, unfortunately we could not find similar calculation methods for ζ_o and ζ_u as $f\left(\frac{D_E}{D}; s_E; A_o; A_u \text{ and draft tube shape}\right)$. Thus we were forced to perform experiments, and as we saw two different possibilities to determine ζ_U resp. ζ_U^*, we actually realized both in two parallel experimental ways[3, 9].

- In one way the circulation numbers n_U were determined by measuring the flow velocity w_a in the annulus of a complete *model JLR* made of acrylic glass with D = 290 mm and $s = 5$ using drinking water as L-system. Then ζ_U (17) and $\frac{Re_m}{Re_1}$ (18) could be directly calculated. In these experiments we kept constant the following values, which we had found out in preliminary test runs to be optimal (see Fig. 6):

$$L_E \approx 7.5\ D_E; a = 0.1\ D; X_o \equiv 4\ \frac{D_E}{D}\frac{A_o}{D} \approx 0.82; X_u \equiv 4\ \frac{D_E}{D}\frac{A_u}{D} \approx 0.58\ . \quad (20)$$

In the main experiments ζ_U and therewith $\frac{Re_m}{Re_1}$ proved to depend only on $\frac{D_E}{D}$ and the draft tube shape[3].
Over the whole range investigated[3, 8, 9] from

$$\begin{aligned} &2 \times 10^4 \leqslant Re_1 \leqslant 6 \times 10^5 & &0 < \dot{M}_1 \leqslant 1.4\ \mathrm{kg\ s^{-1}} \\ &29 \leqslant \frac{D}{D_1} \leqslant 159 & &0.3 \leqslant \frac{D_E}{D} \leqslant 0.82 \\ &27 \leqslant \frac{D}{s_E} \leqslant 100 & &0.04 \leqslant X_{o,\,u} \leqslant 2.8 \end{aligned} \quad (21)$$

it was found for each given geometry in agreement with fluid dynamic experience

$$\zeta_U \approx const. \quad (22)$$

That means according to (18) for each geometry and fluid system

$$\frac{Re_m}{Re_1} \approx const. \quad (23)$$

and to (17)

$$\frac{D_1}{D} \cdot n_U \approx const.\ . \quad (24)$$

The experimental results from the *model JLR* (D = 290 mm; s = 5) with the various draft tubes shown in Fig. 8 are presented in Fig. 9 for ζ_U and in Fig. 10 for $\frac{Re_m}{Re_1}$. For

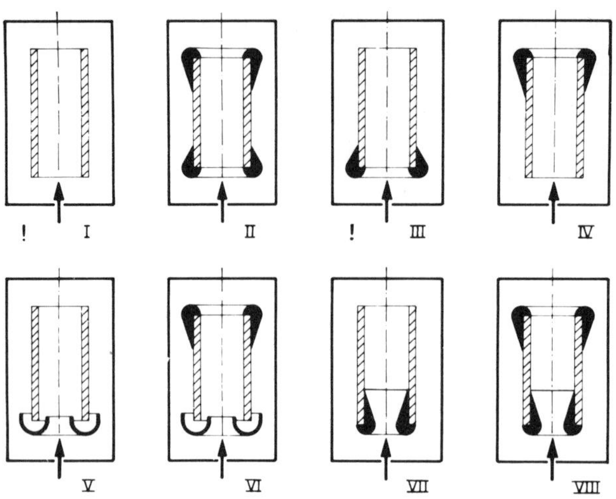

Fig. 8. Types of draft tubes used in model *JLR* (D = 290 mm; s = 5)

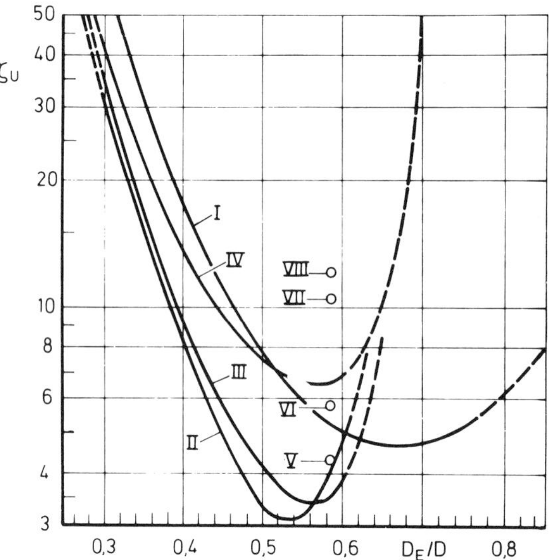

Fig. 9. Circulation resistance number ζ_U in model *JLR* (see Fig. 8)

the smooth cylindrical draft tube (thick curves) it results

$$\left(\frac{\widehat{Re_m}}{Re_1}\right) = 1.03 \text{ at } \left(\frac{D_E}{D}\right)_{opt} = 0.59 \text{ with } (\zeta_U)_{\text{opt}} = 5.0\,. \tag{25}$$

Equation (18) confirms, what the Figs. 9 and 10 show, namely that for the *JLR* $\zeta_U = 4.5$ at $\frac{D_E}{D} = 0.66$ does not coincide with $\left(\frac{\dot{Re_m}}{Re_1}\right)$ because there is still the explicit term $\left(\frac{D_E}{D}\right)^{-1}$.

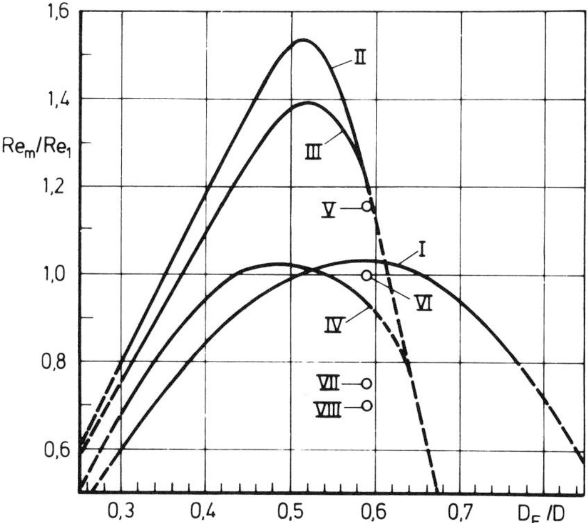

Fig. 10. Relative circulation Reynolds number $\frac{Re_m}{Re_1}$ in model *JLR* (see Fig. 8)

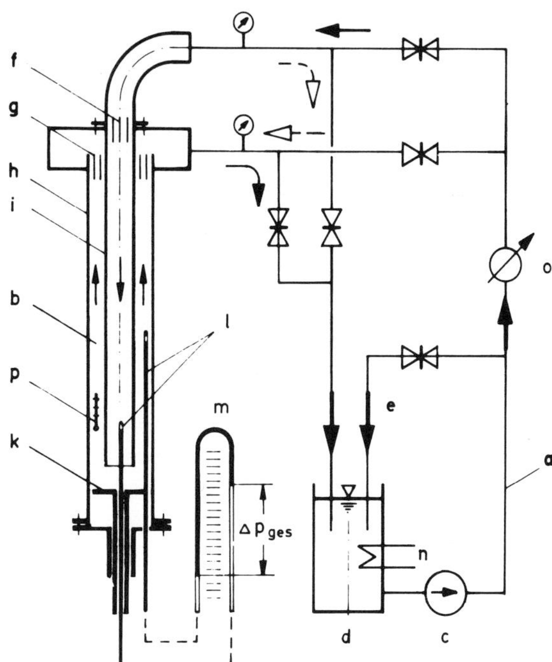

Fig. 11. Scheme of "flow returning apparatus" (D = 140 mm). a liquid circuit, b measuring section, c pump, d storage tank, e flow regulation, f baffle, g baffle, h external tube, i draft tube, k bottom, l pressure gauge, m manometer, n heater, o flow meter, p thermometer

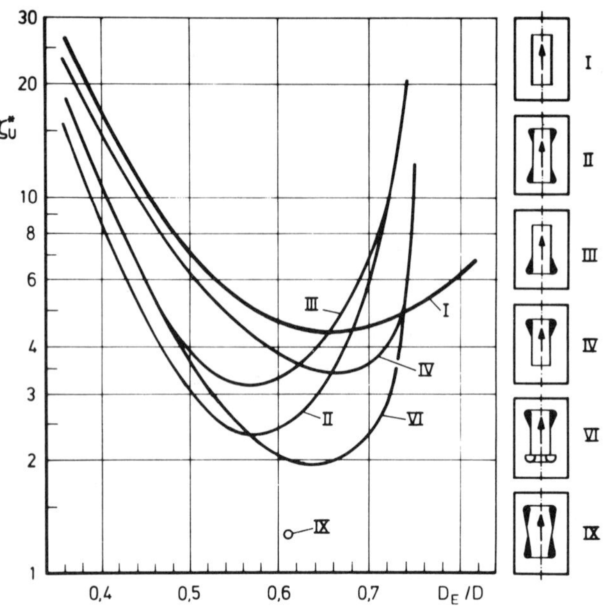

Fig. 12. Circulation resistance number ζ_U^* (19) from flow returning apparatus (Fig. 11)

- In the other series of experiments we measured ζ_o and ζ_u in order to determine ζ_U^* (19). The experimental equipment ("flow returning apparatus") is shown in Fig. 11. Within the external tube h, representing the *LR wall*, the draft tube i is concentrically arranged. An external pumping circuit a produces liquid flows, which can pass the apparatus in both directions. So one can measure the pressure drop caused by the flow returning around the end of the draft tube. Flow direction as indicated in Fig. 11 with black arrows corresponds in Fig. 6 to the flow around the upper end of the draft tube (ζ_o) and the inverse flow direction (white arrows) to the flow around the lower end (ζ_u). In this way ζ_o and ζ_u were experimentally determined and in addition with the calculable ζ_i and ζ_a the circulation resistance number ζ_U^* (19) resulted for the draft tubes as shown in Fig. 12[8], with geometrical and fluid dynamic parameters corresponding to (21).
The ζ_U^*-curves of Fig. 12 determined by aid of the "flow returning apparatus" are in very good agreement with the ζ_U-curves in Fig. 9 as found with the *model JLR*. The results of both investigations lead to exact mutual agreement when a correction factor k_1 is applied to ζ_U in (19), which takes into account the effect of the liquid jet (which only occurs in the *model JLR*) giving

$$\zeta_U = \zeta_i + \zeta_a + \zeta_o + k_1\,\zeta_u\,. \tag{26}$$

The k_1-values resulting from the matching of both sets of experiments are given in Table 1[8].

Table 1

D_E/D	0.3	0.4	0.5	0.6	0.7	0.8
k_1	1.40	1.32	1.24	1.16	1.08	1.00

For the *propeller loop reactor (PLR)* $\zeta_U \equiv \overset{*}{\zeta}_U$ is valid, because there is no liquid jet which can affect ζ_u. Hence at steady state conditions a total circulation flow $\dot{M}_3$ (or $\dot{M}_2$ if $\dot{M}_1 = 0$) results, at which the pressure drop Δp_U due to circulation just equals the pressure head produced by the propeller according to

$$\Delta p_U = \zeta_U \frac{\rho_m}{2} w_m^2 = \rho_m g H_P = p_P \tag{27}$$

Thus for *PLR* one can determine w_m; Re_m; r_U, and $\bar{t}_U$ directly for known values of ρ_m; H_P, and ζ_U. In continuous operation with throughput $\dot{M}_1$ (however in *PLR* not injected as a high velocity liquid jet) n_U and $\bar{t}_U$ have the same meaning for the *PLR* as previously described for the *JLR*.

Figure 12 shows clearly that $\check{\overset{*}{\zeta}}_U$ (and correspondingly $\check{\zeta}_U$) can be lowered from about 4.5 to 1.2 (i.e., to about one quarter) by changing from the cylindrical draft tube I to the conical type IX with deflection bulges, which surprisingly have to be arranged in front of each deflection. According to (18) $\frac{Re_m}{Re_1}$ can thus be doubled!

The ζ_U- and $\overset{*}{\zeta}_U$-values given in Figs. 9 and 12 are valid for a height-to-diameter ratio of $s = \frac{H}{D} = 5$. Then in the *JLR* the momentum of the liquid jet is completely transferred inside the draft tube[10]. Should the loop reactor be designed with higher s, then at constant ζ_o and ζ_u, only ζ_i and ζ_a increase. This can be allowed for by multiplying the $\frac{Re_m}{Re_1}$-values calculated with (18) from ζ_U or $\overset{*}{\zeta}_U$ at $s = 5$ by a correction factor k_s, given in Fig. 13 for other values of s[3, 8, 11].

Now it must be emphasized that in *JLR* equal Re_1 does not however mean equal liquid jet power input P_L.

The volume flow of the injected liquid jet

$$\dot{V}_1 = \frac{\pi}{4} D_1^2 w_1 \tag{28}$$

and its dynamic pressure

$$p_1 = \frac{\rho_1}{2} w_1^2 \tag{29}$$

give its power

$$P_L = \dot{V}_1 p_1 = \frac{\pi}{8} \rho_1 D_1^2 w_1^3 = \frac{\pi}{8} \rho_1 \nu_1^3 Re_1^3 D_1^{-1}. \tag{30}$$

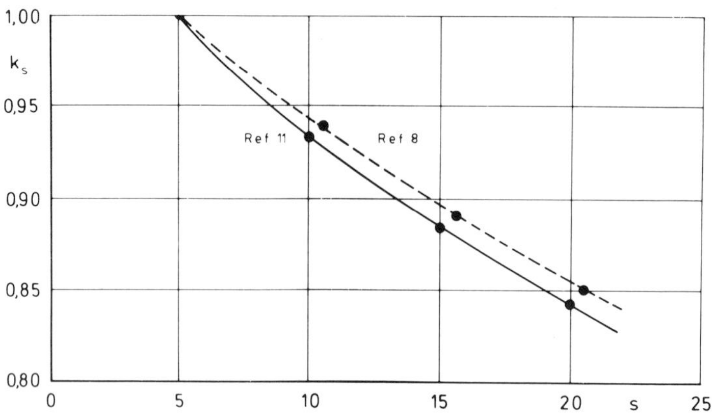

Fig. 13. Correction factor k_s for $\frac{Re_m}{Re_1}$ considering $s \equiv \frac{H}{D} \geqslant 5$

Thus for constant values of ρ_1 and ν_1 it results

$$(P_L)_{\rho_1,\nu_1} \sim Re_1^3 \, D_1^{-1} \tag{31}$$

or

$$(Re_1)_{\rho_1,\nu_1} \sim (P_L \, D_1)^{1/3} \, . \tag{32}$$

Thus the mean circulation velocity w_m – resp. Re_m (11) – can be increased for $\left(\widehat{\frac{Re_m}{Re_1}}\right) \equiv \left(\frac{Re_m}{Re_1}\right)_{opt}$ = const. (23, 25) and P_L = const. by increasing the liquid nozzle diameter D_1 and therewith Re_1 (32)!

This is a very important conclusion for the combining of mixing and dispersing effects in *JLR*, because we shall find, that in a certain range mixing depends on Re_m (84), which is proportional to Re_1 (23), whilst dispersing depends on P_L (129).

These results of theoretical and experimental investigations enable optimal fluid dynamic design of *JLR* and *PLR* for (quasi-)homogeneous aqueous systems with given geometrical, material, and operational conditions. One can base these considerations on ζ_U (17), which was determined (Fig. 9) with the *model-JLR* and thus is valid directly for the *JLR*; or on ζ_U^* (19), which could be achieved (Fig. 12) by means of the flow returning apparatus (Fig. 11) and which is valid directly for the *PLR*. But one can also determine ζ_U (26) with the partial resistance numbers, using k_1 (Table 1) to consider the influence of the L-jet on ζ_u [8].

Both ways of investigation lead to maximal circulation flow for the smooth cylindrical draft tube I – expressed by $\left(\widehat{\frac{Re_m}{Re_1}}\right)$ – at $\left(\frac{D_E}{D}\right)_{opt}$ = 0.59. If one uses other $\frac{D_E}{D}$-relations or draft tube shapes one must take the corresponding values of ζ_U, ζ_U^*, $\frac{Re_m}{Re_1}$ from Figs. 9,

10, and 12. If $s = \frac{H}{D} \neq 5$, one has to multiply the values of $\frac{Re_m}{Re_1}$ with the correction factor k_s, given in Fig. 13 for different s-values.

For *PLR* the required pressure head H_P and pumping power P_P of the propeller results from (27) resp. (40) and (46) to realize w_m at a certain ζ_U^*. For *JLR* the needed power input P_L of the injected L-jet can be calculated with (30). Here it is of special importance, that different values of Re_1 (resp. Re_m) and P_L can be combined by only changing the L-nozzle diameter D_1 (30–32).

It is furthermore important, that for each *LR* with a given geometry it yields $\zeta_U \approx$ const. in accordance with fluid dynamic experiences in other fields. From this fact it can be theoretically derived (18) and experimentally confirmed, that for the *JLR* it must also prove $\frac{Re_m}{Re_1} \approx$ const. (23) independent of D_1 and $\dot{V}_1$. For the *JLR* with aqueous systems and the draft tube type I (Fig. 8) one obtains in good approximation the easy retainable optimal value $\left(\frac{\widehat{Re_m}}{Re_1}\right) \approx 1$ (25).

The validity of the prediscussed results for *JLR* and *PLR* have been confirmed by our experiments with drinking water for reactor diameters D = 20–630 mm and $s = \frac{H}{D} =$ 5–33. Further scale up of D is being studied. From a certain reactor size further enlargement of *JLR* will require other arrangements, e.g., of several liquid nozzles and draft tubes, as indicated in corresponding patents[12].

On the other hand we had of course to investigate the fluid dynamic behaviour of *JLR* and *PLR* with other liquids. These investigations shall only be briefly discussed as far as they may give informations about corresponding aerated biosystems.

3.2 Flow Behavior of PLR and JLR with Highly Viscous (Quasi-)Homogeneous L-Systems

3.2.1 Experimental Fluids

The systematic chemical engineering investigation of aerated highly viscous systems in *LR* is still in the beginning. Therefore only results of experiments with (quasi-)homogeneous highly viscous systems in *PLR* and *JLR* are indicated here[13, 14].

As Newtonian fluids glucose/water mixtures were used with dynamic viscosities of

$\eta = 10^{-3}$ to 5 Pa s for the *PLR*

$\eta = 10^{-3}$ to 0.4 Pa s for the *JLR* .

Notice: 10^{-3} Pa s (≙ 1 cP) is the viscosity of water at $T = 20$ °C.

For Newtonian fluids the well-known relation derived by Newton is valid between the shearing stress τ [N m^{-2} ≡ Pa] and the shear rate $\dot{\gamma} = -\frac{dw_x}{dy}$ [s^{-1}] with the dynamic viscosity η [Pa s]

$$\tau = \eta\,\dot{\gamma}\,. \qquad (33)$$

Carboxymethylcellulose (CMC) in water was used as non-Newtonian fluid. For such pseudoplastic fluids the law of Ostwald and de Waele

$$\tau = K\,\dot{\gamma}^n \tag{34}$$

is exact enough for technical use, where

K: consistency factor
n: flow index .

Analogous to (33), from (34) an apparent viscosity

$$\eta_s \equiv \frac{\tau}{\dot{\gamma}} = K\,\dot{\gamma}^{n-1} \tag{35}$$

can be defined.

$n < 1$ holds for pseudoplastic (structure viscous) fluids, $n > 1$ for dilatant media and $n = 1$ for Newtonian fluids, for which K is identical to the dynamic viscosity η. From Eq. (34) arises, that with increasing $\dot{\gamma}$ the apparent viscosity η_s decreases for pseudoplastic fluids whilst it increases for dilatant media and remains constant for Newtonian fluids with $\eta_s \equiv \eta \equiv K$.

3.2.2 Experimental Equipment

The experimental equipment (with a *JLR*) is shown in Fig. 14, the scheme of the *PLR* in Fig. 15, and the propellers used in Fig. 16.

All *LR* investigated were made of plexiglass and had the following dimensions:
Internal diameter D = 190 mm

Grade of slenderness $s = \frac{H}{D} = 5$ or 10

Diameter ratio $\frac{D_E}{D}$ = 0.67 for *PLR*
= 0.6 for *JLR* .

The speed of the propeller in *PLR* was varied from n_P = 100–1700 rpm and the jet volume flow in *JLR* from $\dot{V}_1 = 0.1-20\,l \cdot s^{-1}$ with nozzle diameters D_1 = 1.8–10 mm.

The following parameters were measured: The circulation volume flow $\dot{V}_P$ (*PLR*) and $\dot{V}_3$ (*JLR*) using the maximal velocity $\hat{w}_a$ in the annulus; furthermore the power input P_P (*PLR*) and P_L (*JLR*).

We shall indicate here only some important results, we found with Newtonian fluids, mainly to demonstrate the influence of viscosity on the flow behavior of *LR* and furthermore to point out similarities between *JLR* and *PLR*, which are of general importance for aerated biosystems too.

3.2.3 Theoretical Relations for Newtonian Fluids

For geometric and hydrodynamic similar designs, characterized by the diameter D_P and the speed n_P of the propeller, the following similarity relationships can be used for the

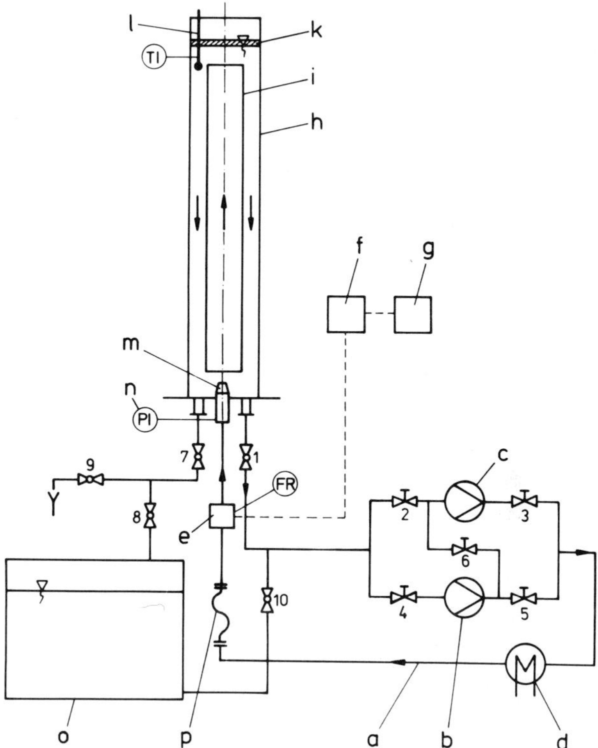

Fig. 14. Scheme of experimental equipment with jet loop reactor *(JLR)* for highly viscous liquids. a liquid circuit, b pump, c pump, d cooler, e flow meter, f impulse amplifier, g frequency counter, h external tube, i draft tube, k cover, l thermometer, m nozzle, n manometer, o storage tank, p high pressure hose

scale-up from models to large-scale units (D_P) or for the change of the speed of propellers (n_P):

Velocities

$$w \sim n_P D_P \,. \tag{36}$$

Cross sections

$$A_P \sim D_P^2 \,. \tag{37}$$

Volume flows

$$\dot{V}_P \sim w A_P \sim n_P D_P^3 \,. \tag{38}$$

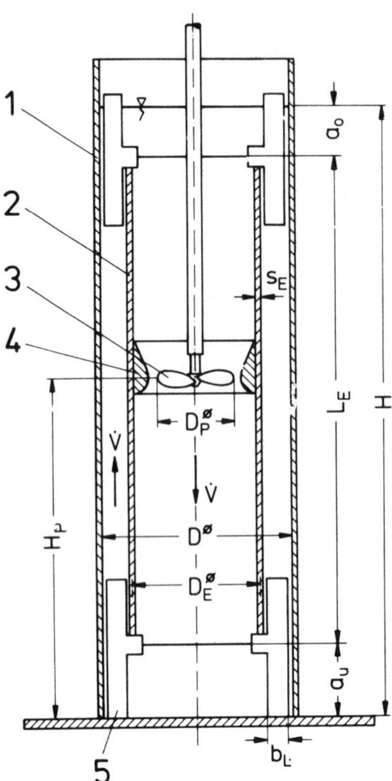

Fig. 15. Scheme of the propeller loop reactor (*PLR*) studied in the experimental equipment of Fig. 14 for highly viscous liquids. 1 external tube, 2 draft tube, 3 propeller, 4 stream profile, 5 baffles

Pressure heads

$$p_P \sim \rho\, w^2 \sim \rho\, n_P^2\, D_P^2\,. \tag{39}$$

Powers

$$P_P \sim \dot{V}_P\, p_P \sim \rho\, n_P^3\, D_P^5\,. \tag{40}$$

Thus the volume flow produced by a propeller – which is in *PLR* the circulation flow – can be expressed as

$$\dot{V}_P = N_V\, n_P\, D_P^3\,. \tag{41}$$

The "volume flow number"

$$N_V \equiv \frac{\dot{V}_P}{n_P\, D_P^3} \tag{42}$$

is a constant characteristic number for a certain propeller, pump or agitator.

Fig. 16. Types of propellers used in *PLR* (Fig. 15)

Furthermore for propellers – as for agitators – the propeller Reynolds number is an important parameter; it is defined using (36) as

$$Re_P \equiv \frac{n_P D_P^2}{\nu} = \frac{n_P D_P^2 \rho}{\eta} \sim \frac{w D_P}{\nu}. \tag{43}$$

For *PLR* the total circulation flow $\dot{V}_P$ – as for *JLR* $\dot{V}_3$ – can be characterized by the mean circulation Reynolds number according to (11) using (42):

$$Re_m \equiv \frac{8 \dot{V}_P}{\nu \pi D} = \frac{8 n_P D_P^3}{\nu \pi D} N_V . \tag{44}$$

Similarly a circulation Reynolds ratio – analogous to that for the *JLR* (12) – can be defined for the *PLR* as

$$\frac{Re_m}{Re_P} \equiv \frac{8}{\pi} \frac{D_P}{D} N_V \sim \frac{D_P}{D} N_V \tag{45}$$

or for constant geometrical parameters

$$\left(\frac{Re_m}{Re_P}\right)_{D_P/D} \sim N_V . \tag{46}$$

Regard the conspicuous simularity between propeller and jet drive as it clearly arises by comparison of (45) with (12) and (46) with (13)!

For propellers (as for agitators) the driving power P_P can be characterized by the "Newton number" according to (40)

$$Ne \equiv \frac{P_P}{\rho\, n_P^3\, D_P^5}\,. \tag{47}$$

For liquid jets one can – as usual in fluid dynamics – relate the excess static pressure P_{L_0} at the nozzle inlet against nozzle outlet to the dynamic pressure $p_1 = \frac{\rho_1}{2} w_1^2$ (29) of the generated liquid jet at the nozzle outlet, thus attaining the "Euler number", which can be transformed with (30) as follows:

$$Eu \equiv \frac{p_{L_0}}{\rho_1\, w_1^2} = \frac{p_{L_0}}{2\, p_1} = \frac{p_{L_0}\, \dot{V}_1}{\rho_1\, w_1^2\, \dot{V}_1} = \frac{P_{L_0}}{2\, P_L} = \frac{4\, P_{L_0}}{\pi\, \rho_1\, w_1^3\, D_1^2} \sim \frac{P_{L_0}}{\rho_1\, w_1^3\, D_1^2}\,. \tag{48}$$

A comparison of (48) with (47) shows, that the Euler number of the injected liquid jet in *JLR* corresponds to the Newton number of the propeller in *PLR*.

One can transform (48) with (30) also to

$$Eu \equiv \frac{P_{L_0}}{2\, P_L} = \frac{4\, P_{L_0}\, D_1}{\pi\, \rho_1\, \nu_1^3\, Re_1^3}\,. \tag{49}$$

The Euler number represents for the liquid nozzle of *JLR* the ratio of the pumping input P_{L_0} (for the prementioned excess static pressure p_{L_0}) to the dynamic liquid jet power output P_L (which is the effective jet power input to the *JLR*).

Thus *Eu* is the reciprocal value of the efficiency of the power transformation within the liquid nozzle from the excess static pressure power P_{L_0} to the dynamic jet power P_L. The smaller the friction losses of this hydraulic power transformation, the more P_L approaches P_{L_0} and *Eu* approaches the value 0.5 (see Fig. 20).

3.2.4 Some Important Results for Newtonian Fluids

a) It is seen from Fig. 17 that for *JLR* in the turbulent range ($Re_1 \geqslant 2000$) $\left(\frac{Re_m}{Re_1}\right)_t \approx 1 \approx const.$ is confirmed, as found in (23) and (25). In the range of higher viscosity $\frac{Re_m}{Re_1}$ decreases as can be expected because in (18) the resistance number ζ_U (14) increases. In the laminar range $\left(\frac{Re_m}{Re_1}\right)_l \sim Re_1$ approximately holds. At higher values of $s = \frac{H}{D}$, ζ_U increases due to greater partial resistances ζ_i and ζ_a (19) and at equal Re_1 consequently $\frac{Re_m}{Re_1}$ decreases (18). The measured points of Fig. 17 can be mathe-

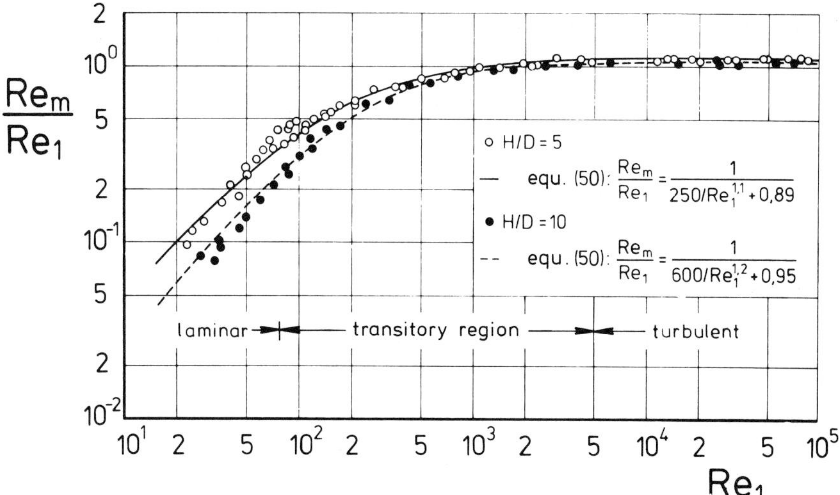

Fig. 17. Circulation Reynolds number ratio $\frac{Re_m}{Re_1}$ of *JLR* as a function of the liquid jet Reynolds number Re_1

matically correlated by the empirical equation (curves)

$$\frac{Re_m}{Re_1} = \left(\frac{C_1}{Re_1^{C_2}} + C_3\right)^{-1} \tag{50}$$

where the constants C_i have to be determined by correlation.

b) A comparison of Figs. 17 and 18 confirms that $N_V \sim \frac{Re_m}{Re_P}$ (46) of a given *PLR* corresponds to $\frac{Re_m}{Re_1}$ (13) of a similar *JLR*.

Thus the measured points in Fig. 18 can be correlated with the same empirical formulation (curves) as for Fig. 17 in (50), namely

$$N_V = \left(\frac{C_1}{Re_P^{C_2}} + C_3\right)^{-1} . \tag{51}$$

From (51) the limiting values are
for low Re_P-values:

$$N_V \approx Re_P^{C_2}\, C_1^{-1} \tag{52}$$

for high Re_P-values:

$$N_V \approx C_3^{-1} \approx const. \, . \tag{53}$$

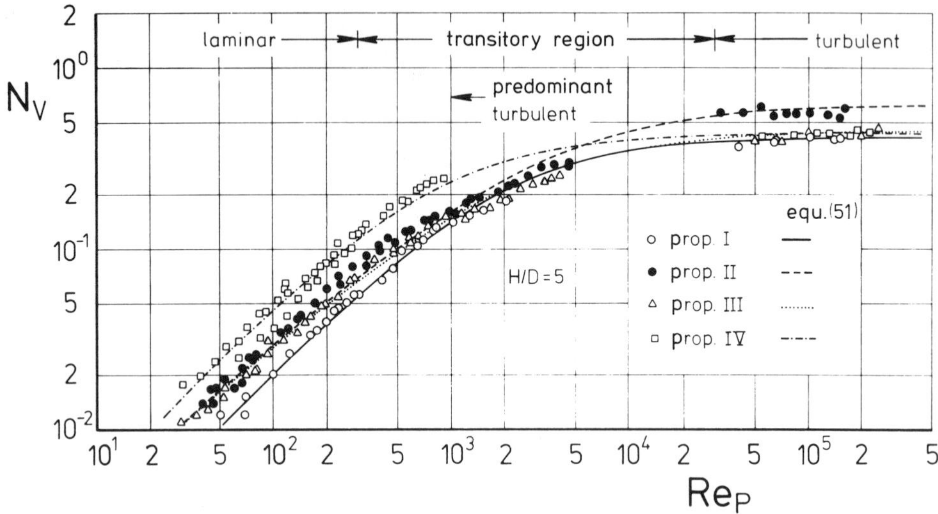

Fig. 18. Volume flow number N_V of *PLR* as a function of the propeller Reynolds number Re_P

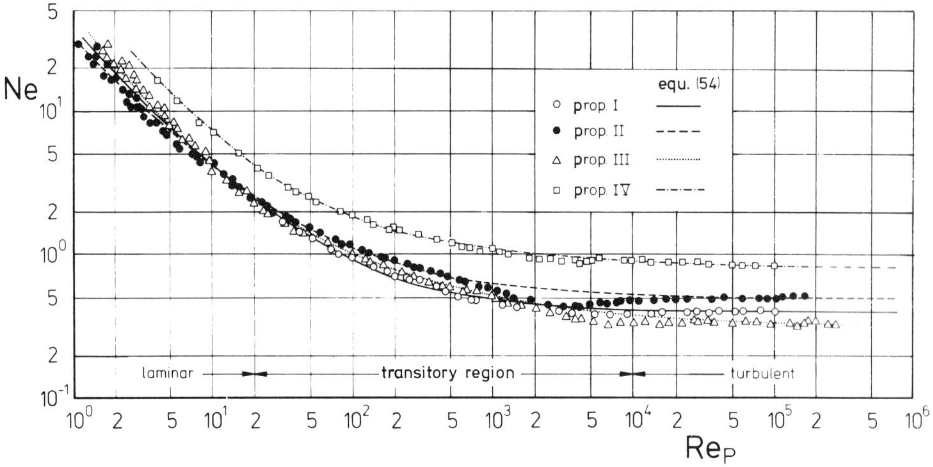

Fig. 19. Newton number *Ne* as a function of the propeller Reynolds number Re_P for different propeller types

c) According to Fig. 19 the course of the Newton number *Ne* over Re_P for the propellers of Fig. 16 corresponds to that of all agitator types investigated[15].
The measured points in this "power chart" can be well correlated by the following empirical equation (curves):

$$Ne = C_1\, Re_P^{-C_2} + C_3\, Re_P^{-C_4} + C_5 \,. \tag{54}$$

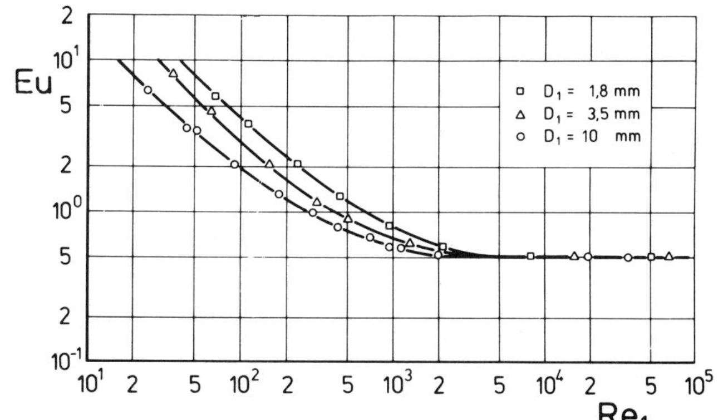

Fig. 20. Euler number Eu as a function of the liquid jet Reynolds number Re_1 for different nozzle diameters D_1

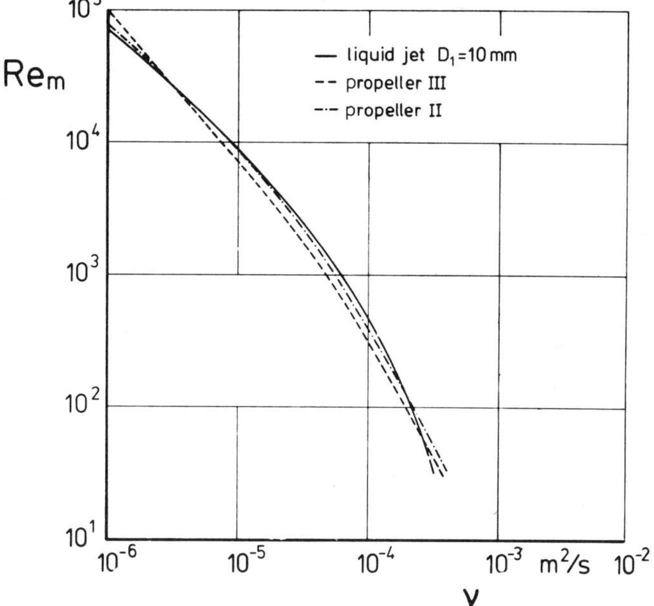

Fig. 21. Comparison of circulation Reynolds numbers Re_m produced by propellers and liquid jets as a function of viscosity ν

d) Figure 20 represents $Eu = f(Re_1)$ for JLR with measured points and calculated curves (49). The similarity to the curves $Ne = f(Re_P)$ for PLR in Fig. 19 is conspicuous. As already remarked to (48), the curves in Fig. 20 approach the value Eu = 0.5 and that here at $Re_1 \gtrsim 4 \times 10^3$. Thus in this range practically holds $P_L = P_{L_0}$ (49). With decreasing Re_1 (i.e., increasing viscosity at constant D_1 and $\dot{V}_1$) Eu increases up to approximately $Eu \sim Re_1^{-1}$. Besides this, Eu is the larger at $Re_1 = const.$,

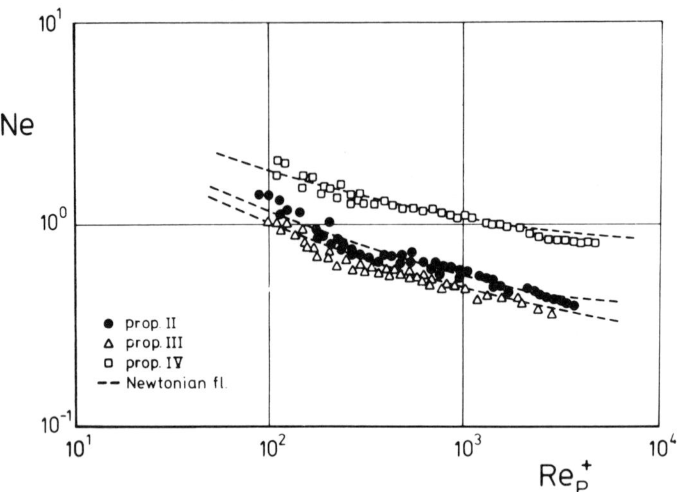

Fig. 22. Newton number Ne as a function of the modified propeller Reynolds number $\overset{+}{Re_P}$ for pseudoplastic liquids

the smaller D_1. For example at $Re_1 = 10^2$ one finds in Fig. 20 with $D_1 = 10$ mm $Eu \approx 2$ or $P_{L_0} \approx 4\, P_L$ (49) and with $D_1 = 1.8$ mm $Eu \approx 4$ or $P_{L_0} \approx 8\, P_L$.

e) Figure 21 shows a comparison of *PLR* and *JLR* with respect to the circulation flow achieved (Re_m) at equal effective power input of $\frac{P}{V_R} = 0.4$ kW m^{-3} as a function of the viscosity. That type of *LR* is the best one concerning the generation of circulation flow, which gives the largest Re_m with the same specific power input and ν = const. Both types of flow drive are surprisingly close together over the wide viscosity range of $10^{-6} \leqslant \nu \leqslant 3 \times 10^{-4}$ m^2 s^{-1}.

3.2.5 Power Demand and Circulation in PLR for Non-Newtonian (Pseudoplastic) Fluids

For non-Newtonian fluids the *apparent* viscosity η_s depends on the shear rate $\dot{\gamma}$ (35). We confine here to the *PLR* using the relation empirically found for agitators[16)]

$$\bar{\dot{\gamma}} = k\, n_P \tag{55}$$

where k is a constant for a certain agitator type. This leads with the apparent viscosity (35) to a modified Reynolds number of the propeller according to (43):

$$\overset{+}{Re_P} = \frac{n_P\, D_P^2\, \rho}{K\,(k\, n_P)^{n-1}} = \frac{n_P^{2-n}\, D_P^2\, \rho}{K\, k^{n-1}}\,. \tag{56}$$

For Newtonian fluids (n = 1) $\overset{+}{Re_P}$ is identical with Re_P (43), where K is then equal to η.

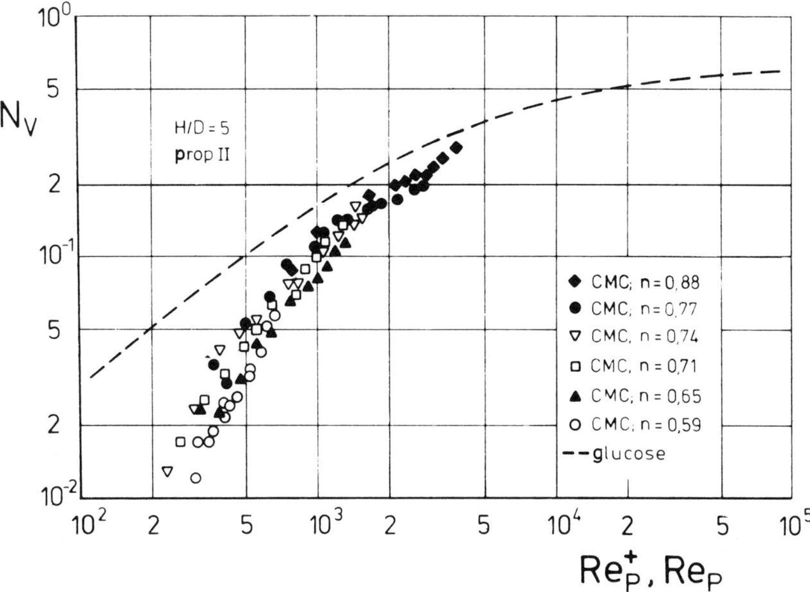

Fig. 23. Circulation volume flow number N_V as a function of the modified propeller Reynolds number $\overset{+}{Re_P}$ for pseudoplastic liquids

As described in[13, 14] one can derive for non-Newtonian fluids "power charts" according to Fig. 22, in which the dashed curves represent Newtonian fluids of Fig. 19. The very good correlation for all propellers confirms the calculation method, as given in[13, 14].

The dependence of the volume flow number N_V (42) on $\overset{+}{Re_P}$ (56) measured with *CMC* deviates however significantly from that of Newtonian fluids, as shown in Fig. 23. This is plausible, because $\dot{\gamma}$ of the total circulation flow is much lower than in the propeller range (determining $\overset{+}{Re_P}$), consequently η_s of the circulation is higher than in the propeller range and thus reduces $\dot{V}_P$ and therewith N_V (42). For increasing flow index *n* (35) Fig. 23 shows the expected approach to the Newtonian glucose solutions ($n = 1$).

Corresponding relationships for *JLR* are discussed in[13, 14]!

The dimensionless relations derived and presented here for highly viscous (quasi-)-homogeneous systems are experimentally proved with the apparatuses, described in 3.2.2, but should be valid too for geometric and fluid dynamic similar scale-up of *PLR* in a wide range and for *JLR* in a certain range, which we are just trying to find out. At least they allow for such homogeneous systems an approximate prediction of the circulation flow and the required power input and for corresponding (quasi-)homogeneous multi-phase systems qualitative references.

3.3 Flow Behavior of Heterogeneous Solid-Liquid-Systems (S-L-Systems or Suspensions) in JLR

In biosystems it is often essential to avoid a settling (sedimentation) of particles, such as cell agglomerates, on the bottom of the reactor. Normally the solid particles should be distributed as uniformly as possible throughout the reaction space.

Therefore the aim of our experiments[17] was to investigate theoretically and experimentally the fluidization and distribution of solid particles as well as the fluid dynamics of suspensions in *JLR*. The plexiglass experimental reactor with $D = 140$ mm is shown schematically in Fig. 24. The material and geometric parameters of the solid particles were varied according to Table 2. Their concentration c_a in the annulus was measured, using a newly developed photometric method incorporating a He-Ne gas laser[17].

Unfortunately it restricts the volume concentration c_{R_o} at uniform distributions throughout the reactor volume V_R of $c_{R_o} \lesssim 1\%$.

The liquid flow $\dot{V}_{L1}$ (Fig. 24) injected at the bottom is removed here at the top, because an outlet at the bottom as in Fig. 6 would cause a removal of solid particles. Therefore, in this case the mean circulation velocity of the liquid is defined correspond-

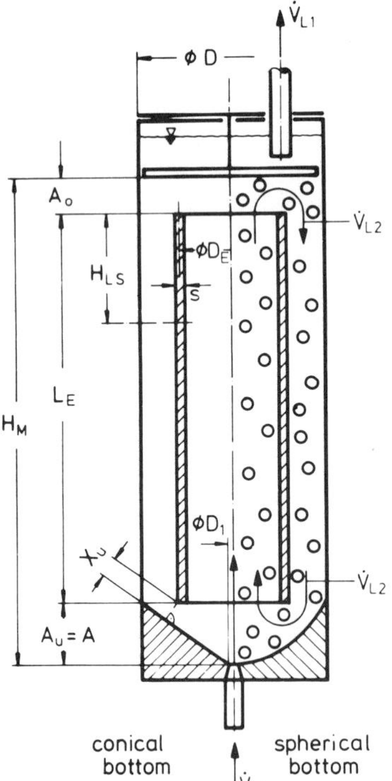

Fig. 24. Schematic figure of the model jet loop reactor (*JLR*) for heterogeneous *L–S*-systems (suspensions); $D = 140$ mm

Table 2

Serial number	Material	Density [g/cm³]	Particle form	Measurements [mm]	Tolerance [μm]
1	PA[a]	1.13	Pellet	Diameter	
2	POM[b]	1.45	Pellet	1.50/2.50/3.175/4.00	20
3	Aluminium	2.70	Pellet		
4	PVC	1.41	Cube	Edge length 4 x 2.5 x 1.5	

[a] PA ≙ Polyamide.
[b] POM ≙ Polyoximethane (Hostaform).

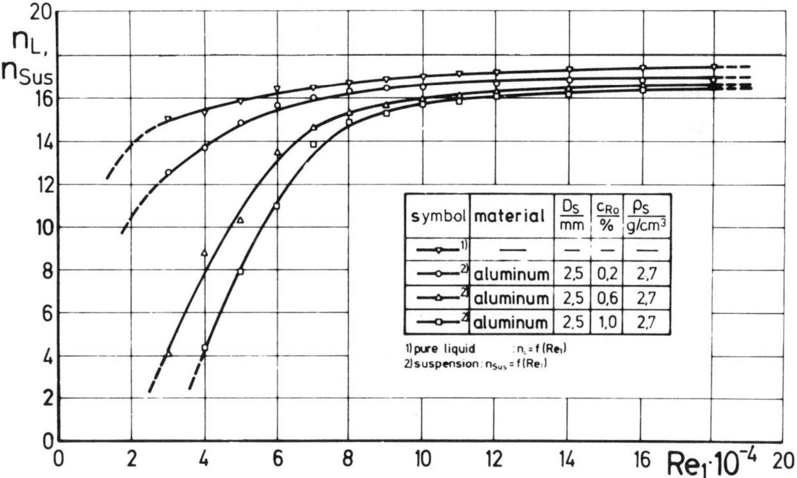

Fig. 25. Circulation numbers n_L and n_{Sus} for various solid concentrations as a function of liquid jet Reynolds number Re_1 $\left(\frac{D_1}{D} = 0.029; \frac{D_E}{D} = 0.69; \frac{L_E}{D} = 4.21; \textit{conical bottom}\right)$

ing to w_m (5) as

$$w_{L_m} \equiv \frac{4\,(\dot{V}_{L1} + 2\,\dot{V}_{L2})}{\pi D^2} \tag{57}$$

and the mean rotation number of the *liquid* corresponding to n_u (4) as

$$n_L \equiv \frac{\dot{V}_{L2}}{\dot{V}_{L1}}\,. \tag{58}$$

The best shape of the bottom was found to be a cone. When the semi-cone angle is larger than the angle of slope of the particle bed, the particles slide on their own into the intake area of the liquid jet.

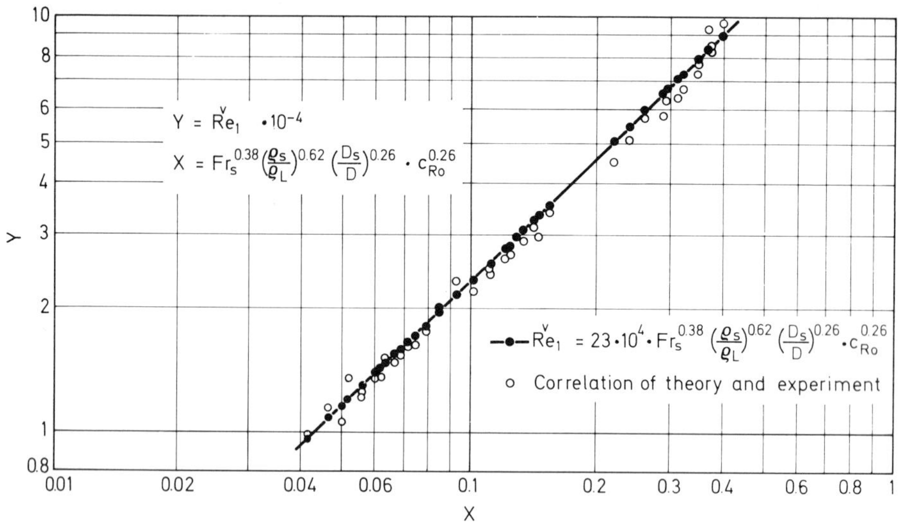

Fig. 26. Minimum Reynolds number of liquid jet $\check{Re}_1$ for rather uniform distribution of solid particles in the whole liquid phase as a function of solid concentration c_{R_0} and solid parameters (δ_S, D_S) $\left(\frac{D_1}{D} = 0.029; \frac{D_E}{D} = 0.69; \frac{L_E}{D} = 4.21; \textit{conical bottom}\right)$

The upper curve of Fig. 25 confirms again for pure liquid flow that according to (12) and (23) $n_L \frac{D_1}{D} \approx const.$ at $Re_1 \gtrsim 10^5$ in this case $n_L \approx 17$ for $\frac{D_1}{D} \approx 0.029$.

For the case of particle loaded flow (suspension) the circulation number n_{Sus} at equal Re_1 is lower than n_L, because part of the liquid jet power input is used to fluidize the solid particles. In[17] one can find experimental and theoretical results for different distributions of the solid particles within the fluid. Here we are mostly interested to know the minimum condition for rather uniform distribution throughout the *whole* reaction space.

It was formulated as

$$\check{Re}_1 = 2.3 \text{ x } 10^5 \; Fr_S^{0.38} \left(\frac{\rho_S}{\rho_L}\right)^{0.62} \left(\frac{D_S}{D}\right)^{0.26} c_{R_o}^{0.26} \tag{59}$$

and presented in Fig. 26.

Here the Froude number Fr_S is defined with the settling rate w_S and the diameter D_S of the solid particles and with the acceleration due to gravity g as

$$Fr_S \equiv \frac{w_S^2}{g D_S} \, . \tag{60}$$

The relations found for the fluid dynamics of suspensions in *JLR* are surely also valid in the dimensionless form for a certain scale-up, which however still must be proved. It is aimed to extend this work especially to cover higher solid concentrations and densities as well as very small particle sizes with respect to microorganisms in biosystems.

3.4 Flow Behavior of Heterogeneous Gas-Liquid-Systems (G–L-Systems) in JLR and ALR

3.4.1 Air-Water-System with JLR and ALR at $w_G \lesssim 10$ cm s^{-1}

The experimental apparatus used for these investigations[18)] is presented in Fig. 27, here with a *JLR* and ring-nozzle sparger at the bottom as Fig. 28a shows in detail. For all plexiglass reactors the dimensions were according to Fig. 7

- internal diameter of the reactor $D = 290$ mm
- diameter ratio $\frac{D_E}{D} = 0.59$
- height-to-diameter ratio of the reaction space $s = \frac{H}{D} = 6 - 22$.

Although several types of spargers, mainly those of Fig. 28, were investigated, the present discussion will be limited to the concentric *ring nozzle* as shown in Figs. 7, 27, and 28a.

Figure 29 presents the characteristic *"flow chart"* of heterogeneous *G–L*-systems in *JLR*. It describes the dependence of the circulation Reynolds number Re_m (for liquid

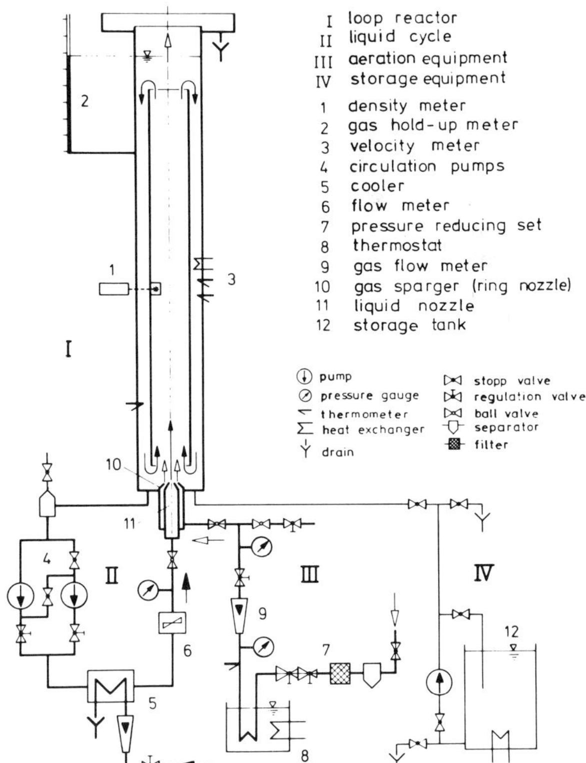

Fig. 27. Scheme of experimental loop reactor plant for *water-air* system; $D = 290$ mm

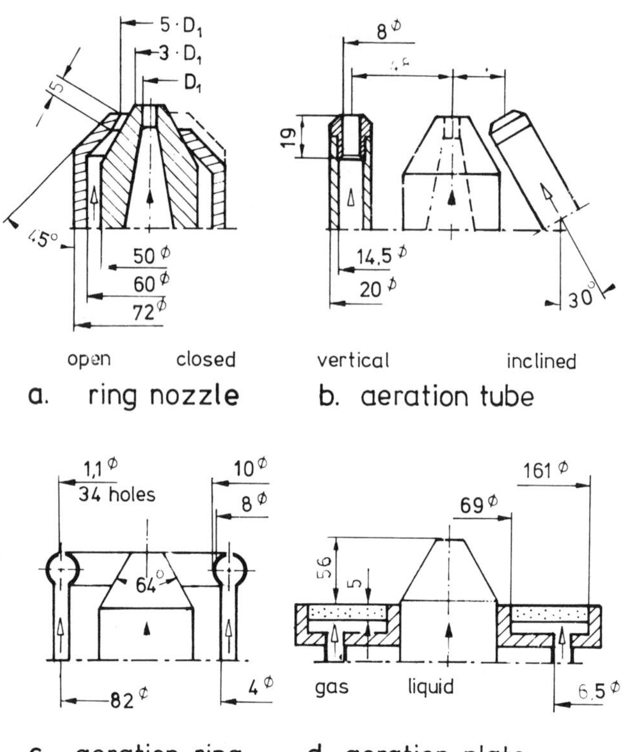

Fig. 28. Gas spargers used in our investigation with G–L-systems

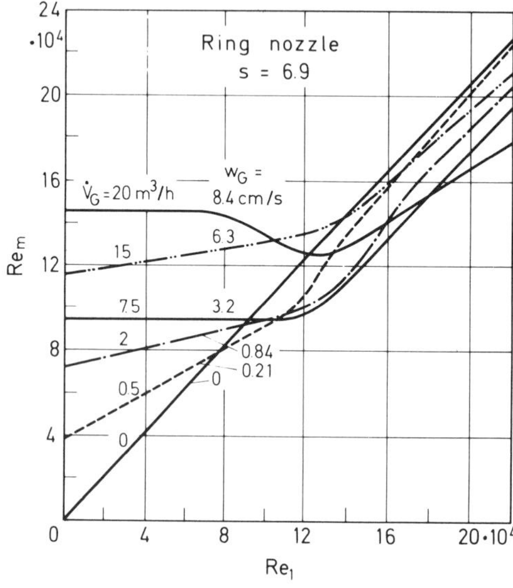

Fig. 29. Liquid circulation Reynolds number Re_m (62) in ALR and JLR with G–L-systems depending on gas throughput $\dot{V}_G$ and liquid nozzle Reynolds number Re_1

circulation only) on the nozzle Reynolds number Re_1 (liquid jet) and on the gas throughput $\dot{V}_G$ or the superficial gas velocity, which refers $\dot{V}_G$ to the *total* reactor cross section

$$w_G \equiv \frac{4\,\dot{V}_G}{\pi D^2}. \tag{61}$$

Re_m is defined here corresponding to (11) regarding only the liquid part $\dot{V}_L$ of the heterogeneous G–L-circulation flow

$$Re_m \equiv \frac{w_{Lm}\,D}{\nu_L} = \frac{8\,\dot{V}_L}{\pi\,\nu_L\,(1-\epsilon)\,D} \tag{62}$$

with the total volumetric gas hold up of V_G in V_R (see Fig. 7)

$$\epsilon \equiv \frac{V_G}{V_R} = \frac{V_G}{V_G + V_L} \approx \frac{H_G}{H} \tag{63}$$

and the mean liquid circulation velocity w_{Lm}, which refers $\dot{V}_L$ to the part of the reactor cross section, which is on the average available for the liquid flow

$$w_{Lm} \equiv \frac{8\,\dot{V}_L}{\pi D^2\,(1-\epsilon)}. \tag{64}$$

From Fig. 29 it follows that:

- Operation with a liquid jet but without gas sparging ($\dot{V}_G = 0$), i.e., *JLR* with homogeneous L-system, is characterized by a linear correlation, which confirms again the previously determined Eqs. (23, 25)

$$\frac{Re_m}{Re_1} = 1.03 = const.$$

- Operation with gas sparging through the ring nozzle, but without a liquid jet ($Re_1 = 0$), i.e., *ALR* with G–L-system, is characterized by the operating points on the ordinate.
- In operation with gas sparging and liquid jet, i.e., *JLR* with G–L-system, 2 ranges can be clearly distinguished:
  - $Re_1 \lesssim 1.2 \times 10^5$ in which the circulation is mainly due to air-lift drive. Nevertheless at low gas throughputs ($w_G \lesssim 1.5$ cm s^{-1}) the liquid jet also contributes to circulation drive, and that increasingly with decreasing w_G. But the jet drive remains smaller than at $Re_1 \gtrsim 1.2 \times 10^5$ (flatter slope of curves).
  - $Re_1 \gtrsim 1.2 \times 10^5$ in which the circulation is mainly due to liquid jet drive.

From this one can conclude that in G–L-systems the liquid jet induces an effective circulation, characterized by Re_m, only at $Re_1 \gtrsim 1.2 \times 10^5$; but even then Re_m is always smaller (up to about 25%) than in homogeneous L-systems. This is mainly due to

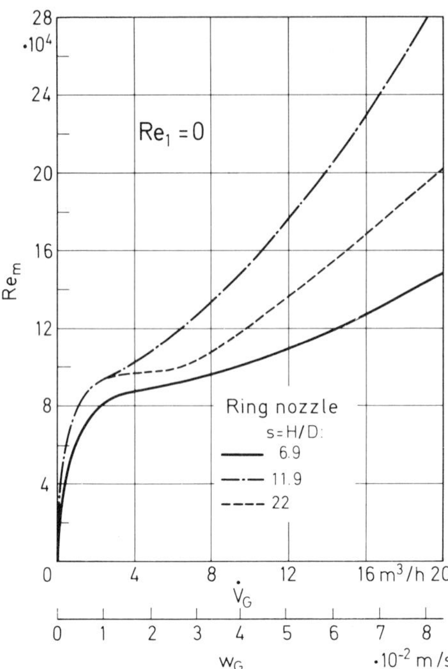

Fig. 30. Liquid circulation Reynolds number Re_m (62) in ALR depending on $(s)_D \sim H$ and $\dot{V}_G$

the fact, that now a considerable part of the L-jet power input is used for gas dispersion, thereby lowering the driving effect.

This is most noticeable when gas is introduced through the ring nozzle shown in Fig. 28a. Then the gas surrounds the liquid jet as a closed envelope and lowers round about its momentum transfer to the liquid phase. This explains why the one-sided sparging with the inclined tube shown in Fig. 28b results in faster circulation at equal power input. Simplified, one can imagine a separation of functions around the L-jet:

The sparged side effects primarily gas dispersion, the other side primarily circulation drive.

Such a sparger is thus not only simpler in construction, but also more effective with regard to dispersion and circulation drive than the ring nozzle (see Sect. 6.3).

Figure 30 shows clearly that the circulation flow of G–L-systems in ALR ($Re_1 = 0$) rapidly increases with increasing height – that means at constant diameter D with increasing grade of slenderness $s = \frac{H}{D}$. This is plausible since the air-lift driving force F_{ALR} increases directly with s, because it is proportional to the height H of the communicating spaces and to the density difference $\Delta\rho$ therein, thus $F_{ALR} \sim H \cdot \Delta\rho \sim s \cdot \Delta\rho$. Certainly the circulation resistance number ζ_U^* (19) increases too with increasing s (at D = const.), but less than proportional to s because only the partial resistance numbers ζ_i and ζ_a increase proportional to s, whilst ζ_o and ζ_u remain constant.

Furthermore Fig. 30 demonstrates that also increasing $\dot{V}_G$ intensifies the air-lift driving force at constant s, for it causes greater differences of local gas hold up inside and around the draft tube and thus greater driving density difference $\Delta\rho$ and therewith driving force $F_{ALR} \sim \Delta\rho$, as mentioned before.

These interrelations between geometric and operational parameters are very important for the design of *ALR* for *G–L*-systems!

3.4.2 Sulfite-System with ALR at $w_G \lesssim 60$ cm s^{-1}

As can be seen from Fig. 29 with increasing gas flow rate $\dot{V}_G$ (w_G) the air-lift drive ($Re_1 = 0$) increases rapidly (ordinate values), whereas the jet drive is practically ineffective at $w_G \gtrsim 8$ cm s^{-1} in the range of $Re_1 \lesssim 1.2 \times 10^5$.

Thus the question arises how the circulation may behave at a much higher $\dot{V}_G$ (w_G) in the *ALR* and *JLR* and how it can be influenced. Since these investigations are still in an early stage, only a few important qualitative results of pre-investigations will be mentioned here for the *ALR*. They could be of especial interest for increasing $\dot{V}_G$-values[19].

These pre-investigations were carried out in the experimental apparatus as shown in principle in Fig. 27. The following dimensions were kept constant (cf. Fig. 7),

$$D = 290 \text{ mm}, \quad A_u = 7 \text{ cm}, \quad L_E = 3.02 \text{ m} .$$

The following parameters were varied,

$$A_o = 10; 39; 100 \text{ cm}$$

$$H = A_u + L_E + A_o = 3.19; 3.48; 4.09 \text{ m}$$

$$s = \frac{H}{D} = 11; 12; 14$$

$$V_R = \frac{\pi}{4} s D^3 = 0.22 - 0.25 \text{ m}^3$$

$$\frac{D_E}{D} = 0.59; 0.67 .$$

The following gas spargers were used at the bottom of the *ALR*:

- open tube with free outlet area $A_G = 10.4$ cm^2
- perforated plate with diameter = 105 mm, hole diameter = 2 mm, and a total free outlet area $A_G = 10.05$ cm^2
- metallic sinter plate with diameter = 105 mm; A_G undefined.

The hydrostatic pressure at the gas inlet becomes increasingly important for taller apparatuses, because it determines the compression requirement. In this respect one tries to introduce the gas as high as possible. For this reason the height of the gas inlet above the reactor bottom, H_B, was varied as follows:

$$H_B = 0; 99; 173 \text{ cm} .$$

The main experiments not only have tested the flow behavior, represented by the liquid velocity w_{L_a} in the annulus, but also the gas hold up ϵ and the specific *G–L*-interface $a_L = \frac{A}{V_L}$ for *ALR* and *JLR*, especially with respect to their suitability as bioreactors for the *SCP*-system[19–21].

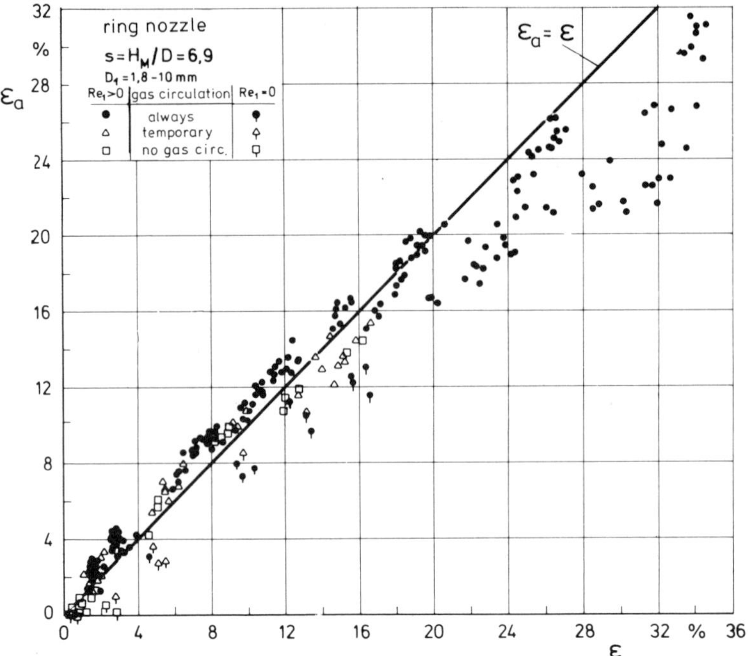

Fig. 31. Local and mean gas hold up for *ALR* (Re_1 = 0) and *JLR* (Re_1 > 0)

Our investigations were carried out with the chemical system of sodium sulphite oxidation to sodium sulphate in aerated aqueous solution with cobalt catalyst, which we call here "sulfite system" with the following overall conversion equation

$$Na_2\,SO_3 + \frac{1}{2}\,O_2 \xrightarrow[Co^{2+}]{k_2} Na_2\,SO_4\ . \tag{65}$$

The specific *G–L*-interface a_R (126) or a_L (127) can be determined for this reaction system according to a chemical method. Of course any transfer and any comparison especially of a_L (127), ϵ (63) or w_{L_m} (64) resp. w_{L_a} (liquid velocity in the annulus) between different systems is problematic indeed, because these parameters very sensitively depend on the coalescence behavior of the *G–L*-systems, and this – as all surface effects – can be strongly influenced even by small changes of components[22–26].

Thus control tests are essential for the transfer of experimental results from model systems and apparatuses to production systems and plants. Nevertheless, model investigations with chemical systems which can relatively simply be handled can provide at least qualitative information for operating behavior of similar bio-systems and design of appropriate biochemical production plants.

In the pre-investigations of *ALR* with sulfite system and high gas flow rates discussed here, the directly measurable parameter w_{L_a}, was selected to determine the circula-

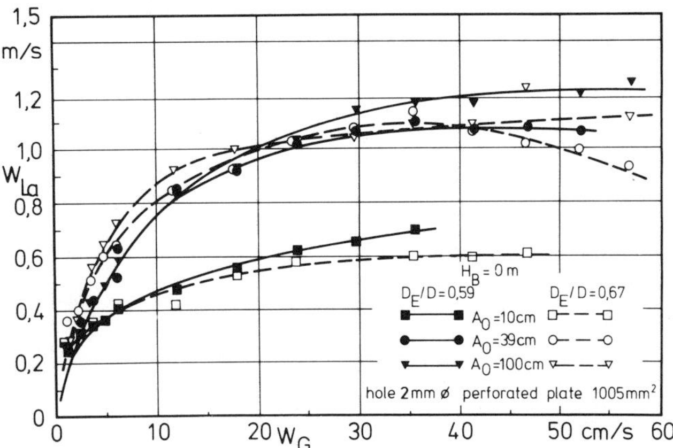

Fig. 32. Liquid velocity w_{La} in annulus; influence of diameter ratio $\frac{D_E}{D}$ and liquid clearance A_o (ALR)

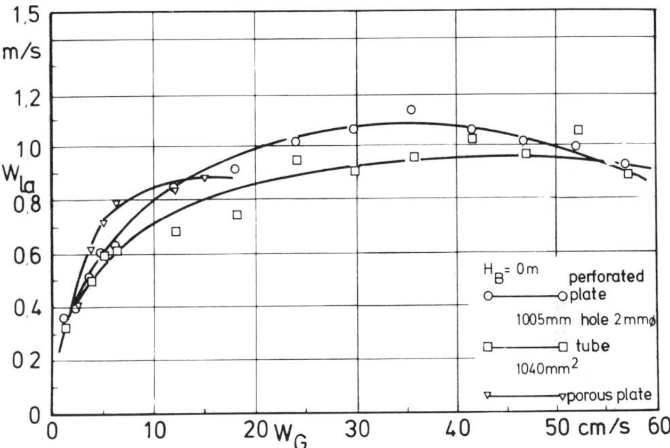

Fig. 33. Influence of gas sparger type on liquid velocity w_{La} in the annulus (ALR)

tion intensity, instead of the mean circulation Reynolds number Re_m (62) of the liquid. The latter is defined by the mean liquid velocity w_{L_m} (64), which depends on the mean total gas hold up ϵ in the reactor space. This is acceptable for the JLR with water-air-system and gas hold up lower than $\epsilon \approx 20\%$, since in this case (cf. Fig. 31) the local gas hold up ϵ_a in the annulus is about the same as the total ϵ[18].

However, this is never the case for an ALR, since its circulation is just based on different local gas contents in the annulus and the internal space. This must be taken into consideration for Re_m as follows,

$$Re_m \equiv \frac{w_{L_m} D}{\nu_L} = \frac{8\, \dot{V}_{L_a} D}{\pi D^2 \nu_L (1-\epsilon)} = \frac{8\, w_{L_a} (1-\epsilon_a) A_a}{\pi D \nu_L (1-\epsilon)} . \tag{66}$$

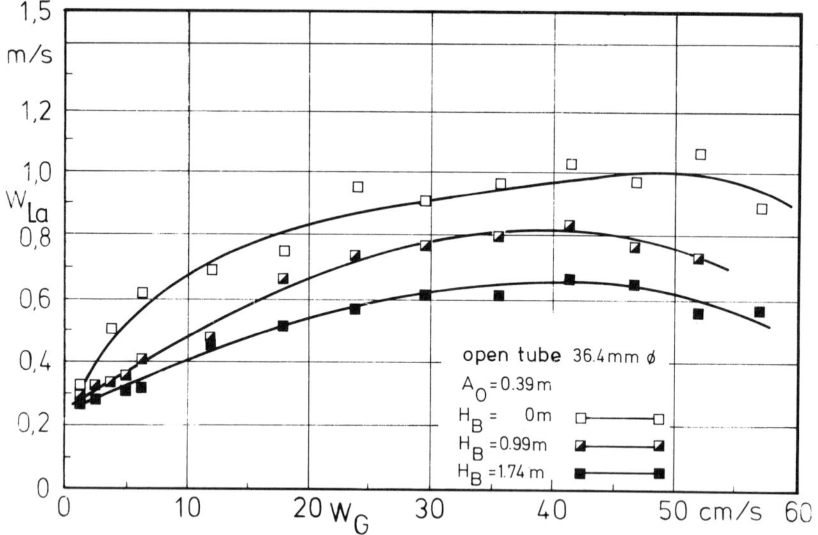

Fig. 34. Influence of aeration height H_B above the reactor bottom on liquid velocity w_{La} in annulus as a function of w_G (ALR)

The value of ϵ_a could not be determined in these pre-investigations, but only ϵ; hence for the present the characterization of the circulation intensity by w_{L_a}!

Figure 32 shows no significant influence of $\frac{D_E}{D}$ on w_{L_a}, but a very strong influence of A_o at $w_G \gtrsim 5$ cm s^{-1}. Thus in this arrangement $A_o \gtrsim 40$ cm should be used, or according to (20):

$$X_o = 4 \frac{D_E}{D} \frac{A_o}{D} \gtrsim 4 . \tag{67}$$

It is also remarkable that at $w_G \gtrsim 40$ cm s^{-1} the liquid circulation can hardly be increased above $w_{L_a} \approx 1.2$ m s^{-1}. Since in this case at $H_B = 0$ (gas inlet at reactor bottom) this gas flow rate with the perforated plate requires already a specific compression power of $\frac{P_G}{V_L} \approx 5$ kW m^{-3}, it is thus not advisable to increase $\dot{V}_G$ furthermore.

The type of gas sparger does not have a significant influence on w_{L_a}, however from Fig. 33 a slight advantage of the perforated plate compared to the tube can be seen for the whole interesting range $w_G \approx 20$–60 cm s^{-1}. The sinter plate should be excluded on account of the very high self-induced pressure drop at $w_G \gtrsim 5$ cm s^{-1}. w_{L_a} decreases rapidly – as Fig. 34 shows for the case of tube sparging – at constant w_G with increasing height H_B of the gas input above the reactor bottom.

This reduction in circulation intensity is plausible, since the air-lift driving force is proportional to the height of the unequally gas-containing communicating spaces; and these are mainly situated *above* the gas inlet.

Additional results of these pre-investigations are discussed in Sect. 4 with respect to gas hold up ϵ, and in Sect. 6.2 with regard to specific interfacial area a_L, to interfacial area A related to power input $\frac{A}{P_G}$ and to specific O_2-input $\dot{m}_{O_2}$.

4 Gas Hold up in LR

4.1 Introduction

For heterogeneous *G–L*-systems the mean volumetric gas hold up ϵ (63) is important mainly for the following reasons:

- the larger ϵ, the larger the mean residence time of gas in the reactor, which is analogous to (8) with (63)

$$\bar{t}_G \equiv \frac{V_G}{\dot{V}_G} = \epsilon \frac{V_R}{\dot{V}_G} \tag{68}$$

thus the larger can be the O_2-conversion of the air and the smaller its volumetric flow rate for a certain specific O_2-requirement. This can lower the operating costs for sterilization and compression of the air

- the smaller ϵ, the larger $V_L = (1 - \epsilon)\, V_R$ (63), in which the reaction occurs, i.e., the better the utilization of the total reactor volume V_R, and thus the lower the investment costs per m^3 of liquid volume V_L
- the larger the local ϵ-difference between the communicating spaces, the more effective the air-lift drive.

The investigations were carried out with water and air[18, 19], or with the previously mentioned sulfite system[19, 27, 28]. Figure 35 shows the principle of the experimental equipment with a *JLR*. If it is operated without liquid jet ($Re_1 = 0$) then we have the *ALR* with ring nozzle sparging.

4.2 Gas Hold up in JLR

The most important results with *JLR* and sulfite system are the following:

Figure 36 shows that the higher Re_1 in *JLR*, the more rapidly ϵ increases, initially linearly with increasing gas flow rate $\dot{V}_G$ (w_G). Then with diminishing increase the curves reach maximum values, which are characterized as "system's gas-loading limit".

If $\dot{V}_G$ is increased above this limit, then ϵ decreases again in *JLR*.

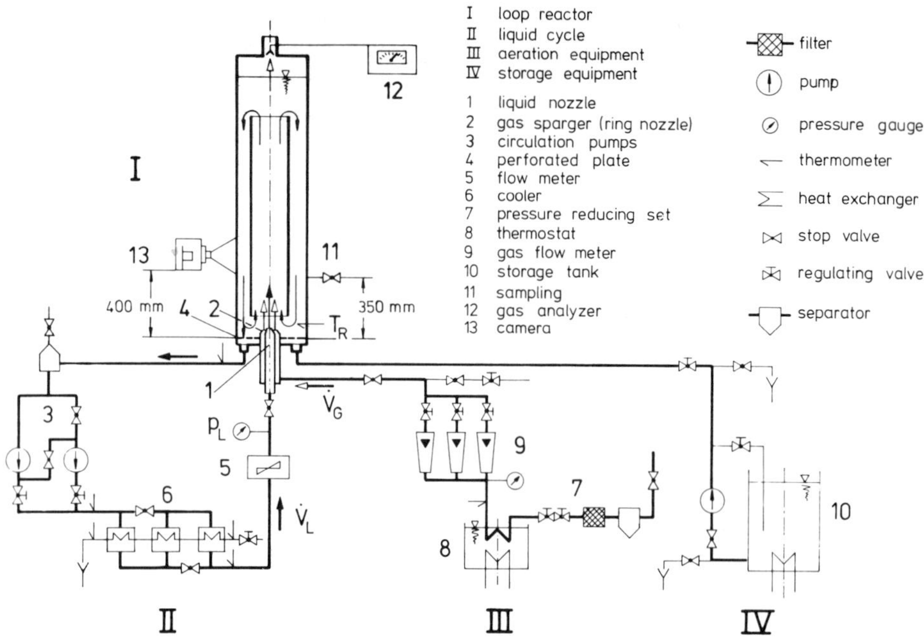

Fig. 35. Scheme of experimental loop reactor plant for *sulfite system* (see Fig. 27 too)

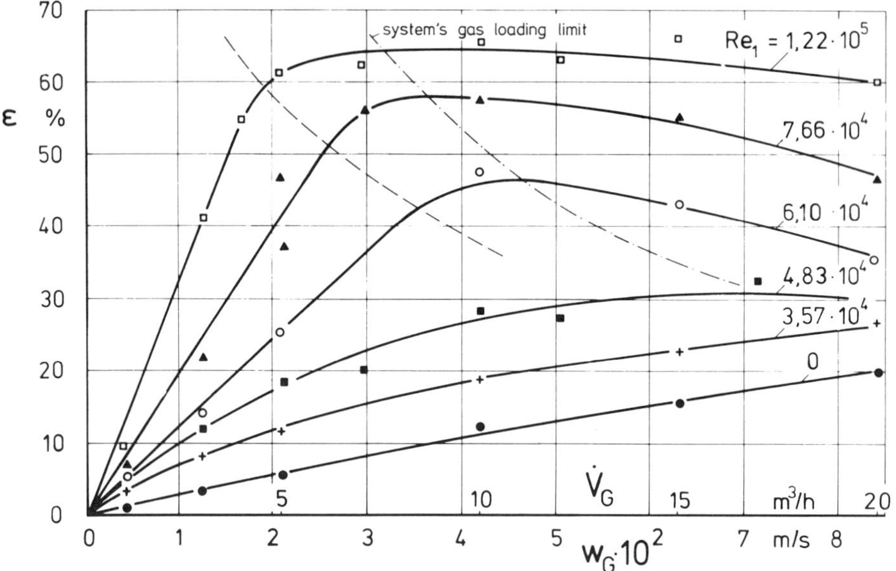

Fig. 36. Gas hold up ϵ in *ALR* (Re_1 = 0) and *JLR* (Re_1 > 0) depending on $\dot{V}_G$ (w_G)

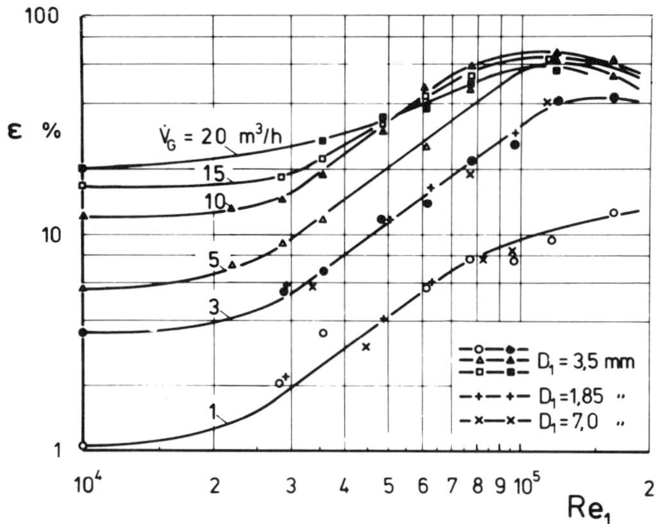

Fig. 37. Gas hold up ϵ in *JLR* as a function of Re_1 for different $\dot{V}_G$

In the case of *JLR*, ϵ clearly depends on Re_1 at $\dot{V}_G = const.$, but not on the specific liquid power input $\frac{P_L}{V_R}$, as one could expect. This is clearly shown by the experimentally determined values for various nozzle diameters D_1 on the lower two curves of Fig. 37. From these experiments it follows in the linear range[28)] the empiric relation

$$\epsilon_{JLR}\ [\%] \approx 2.57 \times 10^{-4}\ w_G\ [\mathrm{m\ s^{-1}}]\, Re_1^{1.4}\ . \tag{69}$$

Substituting (30) in (69) with P_L [kW]; V_R [m³]; D_1 [mm]; w_G [m s⁻¹], one obtains for the *JLR* the numerical equation

$$\epsilon_{JLR}\ [\%] \approx 1.06 \times 10^3\ w_G \left(\frac{P_L}{V_R} D_1\right)^{0.47} . \tag{70}$$

According to (23) the circulation flow, characterized by Re_m, is also a function of Re_1. On the other hand, the dispersion effect – characterized for instance by the specific interfacial area $a_R = \frac{A}{V_R}$ – depends on the specific power input $\frac{P_L}{V_R}$ (129), as will be discussed in Sect. 6.2.

Thus using the relationship (32) $Re_1 = C\,(P_L\, D_1)^{1/3}$ one can combine the wanted dispersion effect (a_R), circulation intensity (Re_m), and gas hold up (ϵ) of the *JLR* by adequate selection of P_L and D_1!

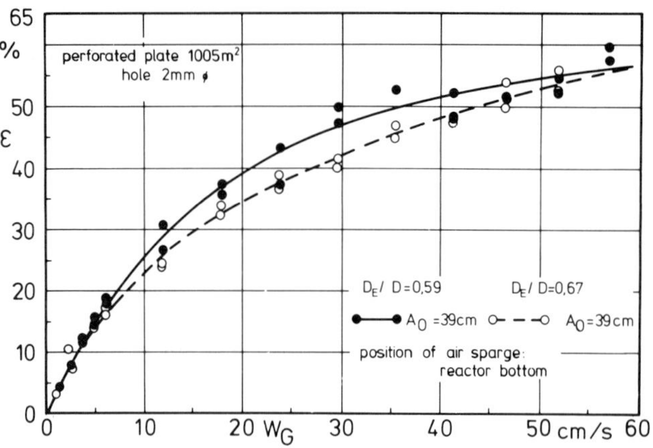

Fig. 38. Gas hold up ϵ in *ALR* as a function of w_G; for different $\frac{D_E}{D}$

4.3 Gas Hold up in ALR

The gas hold up of *ALR* increases in the range $w_G \lesssim 8$ cm s^{-1}, as the lowest curve at $Re_1 = 0$ in Fig. 36 shows, roughly linearly with w_G according to the empiric relation

$$\epsilon_{ALR}\ [\%] \approx 262 \cdot w_G\ [\mathrm{m\ s^{-1}}]\,. \tag{71}$$

With further increase of $\dot{V}_G$ (w_G), ϵ increases less rapidly as Fig. 38 demonstrates[19].

ϵ was always found to be larger at $\frac{D_E}{D} = 0.59$ than at 0.67, and also always larger at $A_o \lesssim 40$ cm than at $A_o = 100$ cm. In the technically interesting range up to $w_G \approx 50$ cm s^{-1} ϵ increases to 50–60%.

5 Mixing and Residence Time Behavior of LR

5.1 Basic Principles

It is often required to distribute all components of a reaction system as rapidly and as uniformly as possible throughout the entire reaction volume, therewith also ensuring a uniform temperature. The required mixing effect should be achieved with as low a power input as possible.

In the *LR* two fundamentally different mixing effects superimpose each other, namely

- *longitudinal mixing* in each circulation and
- *backmixing* due to recycling of the circulation flow $\dot{M}_2$ (Fig. 2).

Longitudinal mixing in the circulation flow is caused, as in real tubular flow, by stream profile, turbulence, dead spaces, and molecular diffusion. As in the case of tubular flow, the resulting longitudinal mixing can be mathematically formulated here according to two models:

- In the *tanks in series model* the real tube (in this case the circulation) is replaced by a series of stirred tank reactors (*STR*) with the equivalent number n_{eq} of consecutive equalvolume ideal *STR*, which results in the same longitudinal mixing effect. With increasing n_{eq}, the real flow approaches the plug flow of the ideal tube reactor ($n_{eq} \to \infty$).
- On the other hand, in the *diffusion model* the prementioned elemental processes – although physically different, but all of essentially statistical nature – are considered according to Fick's laws of molecular diffusion, by summing them all up in the effective longitudinal diffusion coefficient D_{eff}. The diffusion model is characterized by the *Bodenstein number*

$$Bo \equiv \frac{w\,L}{D_{eff}} \tag{72}$$

or applied to *LR* with $L_U = 2\,H$ according to (6) and (7)

$$Bo \equiv \frac{w_m\,L_U}{D_{eff}} = \frac{w_m\,2\,H}{D_{eff}}\,. \tag{73}$$

The experimental determination of Bo is described in[11, 30].
It is a measure of the relationship between mass transport due to plug flow ($w\,L$) and that due to superimposed effective longitudinal mixing (D_{eff}). With increasing Bo, the real flow approaches the plug flow of the ideal tube ($Bo \to \infty$). When $Bo > 8$ both models can be connected by the relation[29]

$$n_{eq} = 1 + \frac{1}{2}\,(Bo^2 + 1)^{1/2} \approx 1 + \frac{Bo}{2}\,. \tag{74}$$

To derive mathematical relations for the calculation of mixing and – in continuous operation – residence time behaviour of *LR* according to the *diffusion model,* we now imagine first that the consecutive circulations in *LR* be transformed to a linear tubular flow, as shown in Fig. 39.

Referring the distance L travelled to the characteristic length $L_U = 2\,H$ of one circulation, we define the relative flow path as

$$x_U \equiv \frac{L}{L_U}\,. \tag{75}$$

If a pulse of tracer M_0 (e.g., dye or electrolyte) is injected to the flow at time $t = 0$ and at the point $x_U = 0$ according to the Dirac impulse function (imaginary infinitely thin disk of tracer), then this will spread along its flow path more and more forward and backward in flow direction due to longitudinal mixing. Imagining oneself as a *stationary ob-*

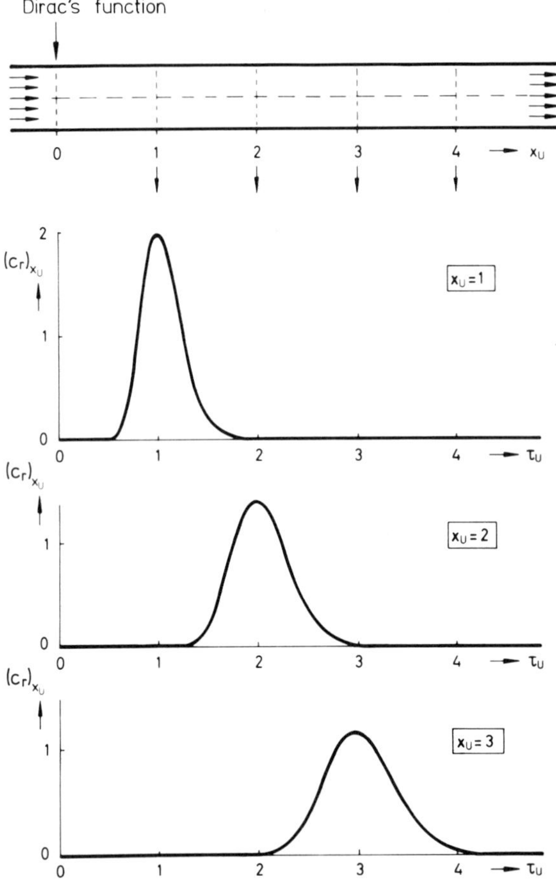

Fig. 39. Impulse response in a real tubular flow characterizing longitudinal mixing

server at the point $x_U = 1$ (corresponding to the end of the first completed circulation), then one measures here a bell-shaped distribution of tracer concentration passing ones place as a function of time, which can be expressed in the dimensionless form $c_r = f(\tau_U)$ represented in Fig. 39.

On the ordinate the relative concentration is plotted, defined as

$$c_r \equiv \frac{c}{c_\infty} \tag{76}$$

with

$$c_\infty \equiv \frac{M_0}{V_R} = \frac{4 \cdot M_0}{\pi \cdot D^2 \cdot H} \tag{77}$$

which is the mean concentration of the added tracer mass M_0, if this were uniformly distributed throughout the entire volume V_R of the *LR*.

The relative time τ_U on the abscissa refers the time t to the mean circulation time $\bar{t}_U$ (7) in the LR, which can be expressed too by the circulation rate r_U (6) or the mean residence time $\bar{t}$ (8) resp. the relative residence time τ (91) and the rotation number n_U (4) as

$$\tau_U \equiv \frac{t}{\bar{t}_U} = t\, r_U = \frac{t}{\bar{t}}\, n_U = \tau\, n_U \equiv i_U\, . \tag{78}$$

Evidently τ_U is identical with the number i_U of mean circulations – with the imaginary mean velocity w_m (5) – within the time t.

Stationary observers, placed in the transformed linear flow at the points $x_U = 1;2;3\ldots$ (corresponding to the ends of the 1st, 2nd, 3rd ... completed mean circulation in LR) observe one after another an increasingly stretched concentration distribution passing their observation point as a function of time t resp. τ_U, as the diagrams $(c_r)_{x_U} = f(\tau_U)$ in Fig. 39 demonstrate.

According to the diffusion model one can derive the local and temporal change of concentration of tracer, injected at $x_U = \tau_U = 0$ as an impulse function into the linear tubular flow of Fig. 39 (above) with Fick's laws of diffusion in the dimensionless form[33)]

$$\frac{\partial c_r}{\partial \tau_U} + \frac{\partial c_r}{\partial x_U} - Bo^{-1}\,\frac{\partial^2 c_r}{\partial x_U^2} = 0\, . \tag{79}$$

With D_{eff} = const., the solution of (79) for the prementioned conditions describes the tracer distribution function of time for each stationary observer at $x_U = 1, 2, 3 \ldots$ in the dimensionless form[32)]

$$(c_r)_{x_U} = \left(\frac{Bo}{4\,\pi\,\tau_U}\right)^{1/2} exp\left[-\frac{(x_U - \tau_U)^2\, Bo}{4\,\tau_U}\right]. \tag{80}$$

These relative concentration functions are represented in Fig. 39 for a tubular flow with a certain Bo, as it is measured by the first three stationary observers at $x_U = 1, 2, 3$. It can be calculated by putting Bo and x_U = const. into (80), and different values of τ_U as long before and after $\tau_U = x_U$ as tracer mass can be measured by the observer at this x_U (see Fig. 39: $x_U = 3$). Example: Constant values of $Bo = 100$ and $x_U = 3$; then put into (80) variable values of $\tau_U = 1.5$–4.5.

To bring backmixing due to recycling into our consideration we now transform the linear tubular flow back to the actual circulation flow of the LR. Then all the above mentioned stationary observers from the points $x_U = 1, 2, 3 \ldots$ come together at one and the same point, namely at the end of the circulation flow, where it returns to its starting point. Let us imagine now that each of them brings along with him his previously (in the tubular flow) measured relative distribution function of time τ_U (80) from Fig. 39 and plots it into one graph, as shown in Fig. 40, against the relative time τ_U. Then the sum of all these inputs must give the total relative concentration c_r passing the joint standpoint of all observers as a function of time. Thus it must be calculable – now for a

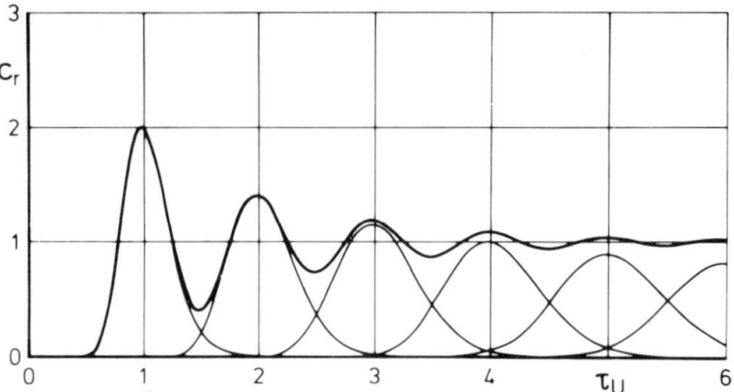

Fig. 40. Impulse response in the real circulation flow of *JLR* (Bo = 50) according to (81)

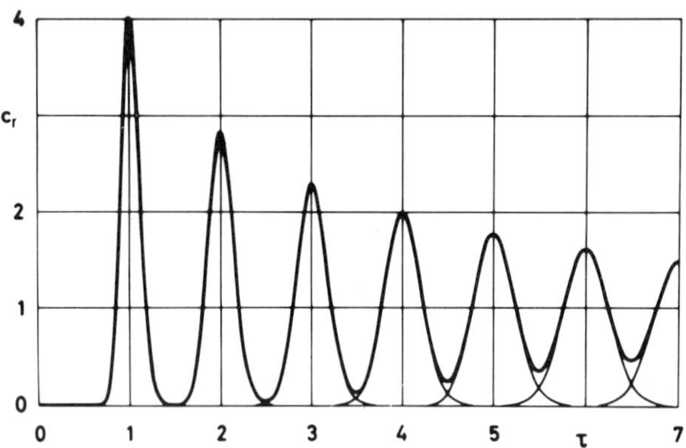

Fig. 41. Impulse response in the real circulation flow of *JLR* (Bo = 200)

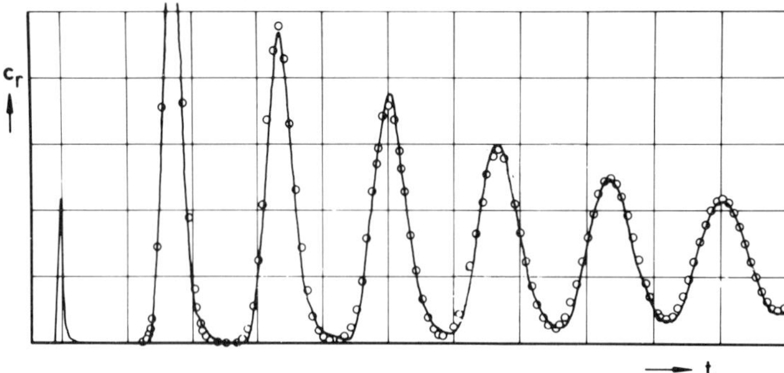

Fig. 42. Measured impulse response curve in *JLR* with s = 15 and Re_m = 4.5 x 10^4 shows good agreement with the points calculated with (81)

certain value of τ_U - by forming the sum of (80) as follows

$$c_r = \Sigma\,(c_r)_{x_U} = \sum_{x_U = \tau_U - 2}^{x_U = \tau_U + 2} \left(\frac{Bo}{4\,\pi\,\tau_U}\right)^{1/2} exp\left[-\frac{(x_U - \tau_U)^2}{4\,\tau_U}\;Bo\right]. \tag{81}$$

The calculations of the distribution functions $(c_r)_{x_U} = f(\tau_U)$ showed, as can be seen too in Fig. 39, that by the stationary observers at each $x_U = 1, 2, 3$ ... considerable concentrations could only be measured in the range of $\tau_U = x_U \pm 2$. Thus summing up of all these $(c_r)_{x_U}$ - functions from Fig. 39 to the total concentration $c_r = f(\tau_U)$ (Fig. 40) we must only take into account for each τ_U the contributions of the individual observers $x_U = \tau_U \pm 2$. The calculation regards only integral numbers of $x_U = 1, 2, 3$ ...; that means, the distribution functions of our imagined stationary observers at the ends of completed circulations. Example: For $\tau_U = 3$ we sum up in (81) the distribution functions (80) for $x_U = 1, 2, 3, 4, 5$. This summing up of the *longitudinal mixing effect,* which we discussed before isolated in the imagined linear tubular flow, represents the characteristic *back-mixing* effect of the *LR* and combines both to the resulting *total mixing* effect of the *real circulation flow* in *LR*.

The curves of Figs. 40–42 were calculated in this way with (81). A comparison of Figs. 40 and 41 shows that with increasing *Bo*, i.e., with approach to the ideal tube reactor, the longitudinal mixing effect in the circulation flow of *LR* decreases. For *LR*, *Bo* = 200 corresponds to a medium longitudinal mixing intensity and *Bo* = 50 to a high one. For a conventional *STR* is $Bo \approx 4$.

Figure 42 confirms the very good agreement between the impulse response measured in *LR* and that calculated pointwise with (81). This equation should, according to its dimensionless formulation, be valid for a certain range of geometrical, material, and operational parameters, which has to be proved by further experiments.

5.2 Degree of Mixing and Mixing Time

Usually mixers are classified according to the mixing time required to achieve a certain degree of mixing throughout the mixer space. Comparisons are only of value when based on the same degree of mixing, which we determine by the inhomogenity as shown in Fig. 43 defined as[30)]

$$h \equiv \hat{h} \equiv \frac{\hat{c} - c_\infty}{c_\infty} = \hat{c}_r - 1\,. \tag{82}$$

In Fig. 43 the maxima and minima of the impulse distribution, calculated with (81) are connected by envelope curves, which are approximately exponential functions. This is proved by Fig. 44, representing the results of calculations for the range of $50 < Bo < 400$, which corresponds to the longitudinal mixing effect of the *LR*, used in our experiments. For the maxima of the impulse distribution the following empiric equation gives a good

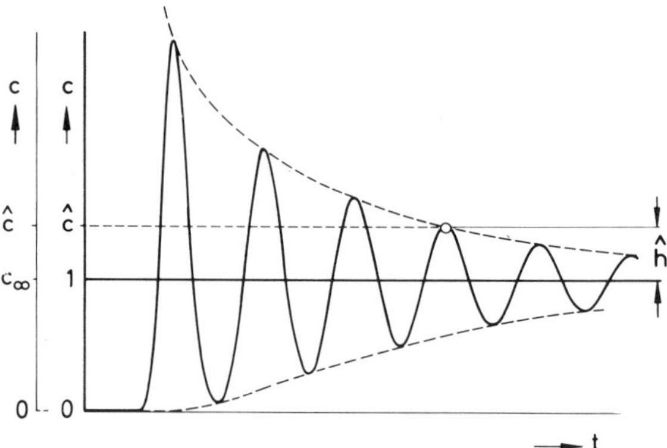

Fig. 43. Determination of the inhomogenity $h \equiv \hat{h}$ characterizing the degree of mixing by the upper envelope curve connecting the maxima of the impulse distribution function

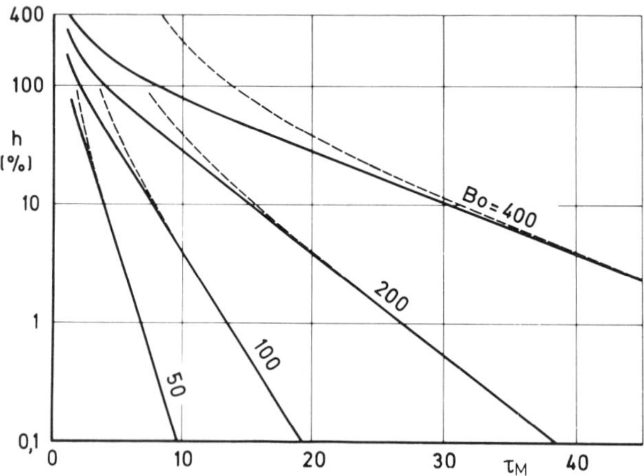

Fig. 44. Inhomogenity h, characterizing the degree of mixing, as a function of the relative mixing time τ_m (88) for different values of Bo (73)

approximation to the experimental results[30)] using the relative mixing time of (88)

$$\tau_M = Bo \, \frac{0.692 - \ln h}{39.48} \, . \tag{83}$$

Now τ_M can be replaced by $t_M = \tau_M \, \overline{t}_U$ (88) and $\overline{t}_U$ by reactor cross sections, volume flows, flow velocities and the flow path $L_U = 2H$ of the circulation in LR. Thus one achieves the wanted equation, to calculate the mixing time required to attain the specified inhomogenity h for LR with $s > 3$ and $h < 50\%$ and the units t_M [s]; D [m]; ν_m [$m^2 \, s^{-1}$]:

$$t_M = \frac{(0.692 - \ln h)(s - 0.24) D^2}{19.74 \, \nu_m} \, \frac{Bo}{Re_m} \, . \tag{84}$$

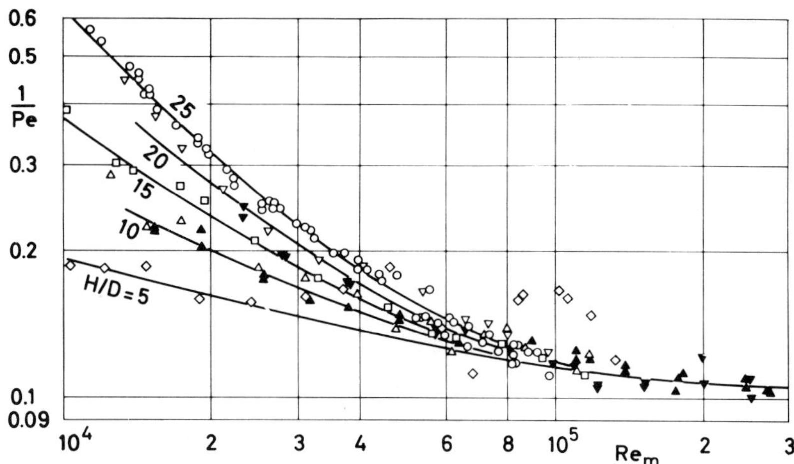

Fig. 45. The longitudinal mixing effect of *LR*, characterized by *Pe* (86) and (87) as a function of Re_m (11) for several values of $s = \frac{H}{D}$

To characterize longitudinal mixing as an "effective diffusion" according to the diffusion model one can use instead of $Bo = \frac{w_m L_U}{D_{eff}} = \frac{w_m 2H}{D_{eff}}$ (73) for the *LR* the well known *Peclet number* for diffusion

$$Pe \equiv \frac{w_m D}{D_{eff}} \equiv Re\,Sc \tag{85}$$

which is analogous to the *Peclet number for heat conduction*

$$Pe_h \equiv \frac{w D}{a} \equiv Re\,Pr\,. \tag{86}$$

Thus to the *LR* evidently applies

$$Bo = 2\frac{H}{D}Pe\,. \tag{87}$$

Fig. 45 states, that *Pe* and therewith *Bo* too are only functions of Re_m at $s = const.$

Thus according to (84) t_M only depends on Re_m, ν_m, D, and s. For a required Re_m, in *JLR* Re_1 is fixed (18) and (23) and thus also the product of jet power P_L and nozzle diameter D_1 (32).

Furthermore it is interesting that t_M increases roughly with D^2 (i.e., $V_R^{2/3}$).

For an inhomogenity $h = 33\%$ the relative mixing time

$$\tau_M \equiv \frac{t_M}{t_U} \tag{88}$$

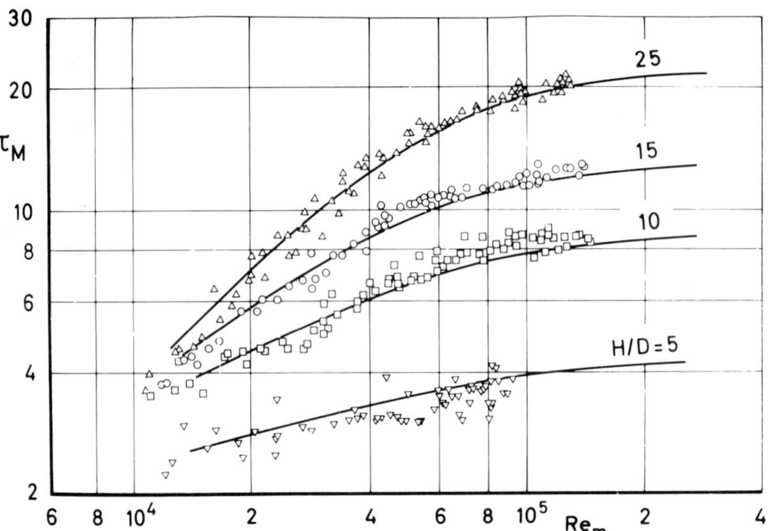

Fig. 46. Relative mixing times τ_M of an *LR* as a function of Re_m (11) for several values of $s = \dfrac{H}{D}$

is presented in Fig. 46; the curves are calculated with (84) and the points are measured values.

It is very important to realize that the mixing time t_M required to achieve a certain inhomogenity h only depends on the variable Re_m – besides the given parameters ν_m; D, and s. According to Fig. 46 the *relative* mixing time τ_M (88) amazingly increases with increasing Re_m, although less than proportional. But one must consider that $\overline{t_U}$ decreases inversely proportional to Re_m (7, 11); in consequence $t_M = \tau_M \overline{t_U}$ (88) decreases with increasing Re_m (84) and in *JLR* thus with increasing Re_1 (23) and D_1 (30) at constant power input P_L. To shorten t_M one therefore needs large $Re_m \sim Re_1 \sim (P_L \cdot D_1)^{1/3}$ (32) and small $s = \dfrac{H}{D}$ (84). A substantial shortening of mixing time t_M can furthermore be achieved by dividing the draft tube of *LR* into 2 or 3 sections, as illustrated in Fig. 47. These mixing times can also be calculated[31]. The results are shown in Fig. 48. It is evident (as already discussed) that t_M increases rapidly with increasing $s = \dfrac{H}{D}$, because the fraction of the tubular flow with relatively small longitudinal mixing effect ($Bo \approx 50$–400) becomes larger compared to the very effective mixing by recirculation. The considerable shortening of the mixing due to divided draft tubes is significant; for example at $s = \dfrac{H}{D} = 10$, t_M decreases from 400 s for a draft tube with one part to 80 s with 2 parts and to 50 s with 3 parts. Reducing $\check{\zeta}_U$ with variously shaped draft tubes, as shown in Figs. 8, 9, and 12, also can shorten the mixing time t_M by increasing $Re_m \sim \check{\zeta}_U^{-1/2}$ (18).

The considerations of the impulse distribution functions and the needed mixing time to achieve a certain degree of mixing concerned hitherto only *LR* with batch op-

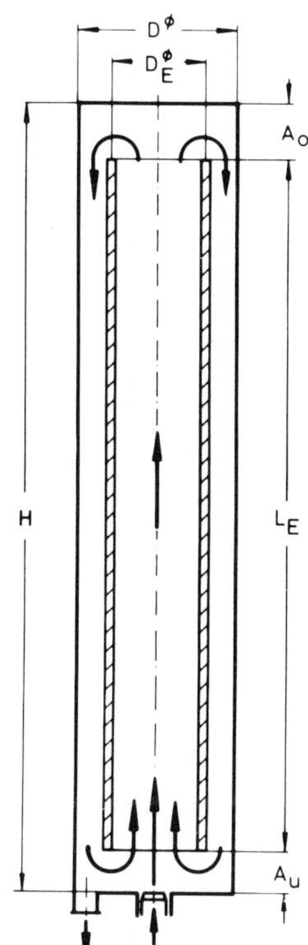

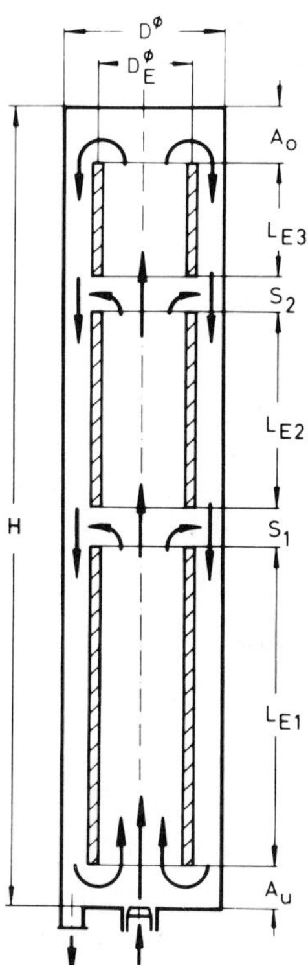

Fig. 47. Types of divided draft tubes for mixing experiments

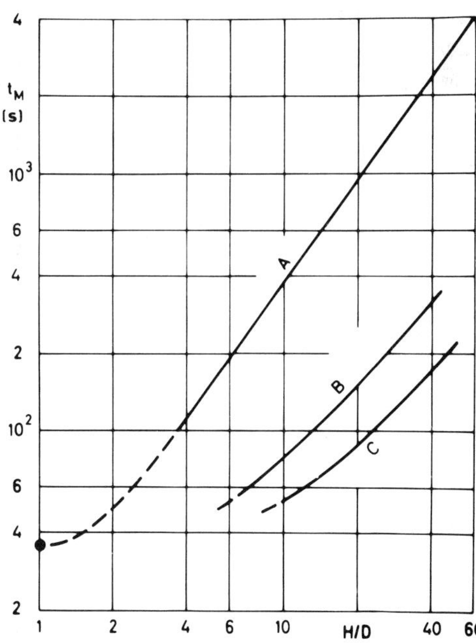

Fig. 48. Calculated mixing times for *LR* with divided draft tubes A: 1 section; B: 2 sections; C: 3 sections

eration. But they can be applied to continuous operation too, as it will arise from the next chapter.

Furthermore the impulse distribution function can be calculated for whatever points of the circulation one chooses, not only those at the end of the circulation. In all cases the calculated impulse distribution functions lead to the same envelope curves of the maxima and minima provided that $D_{eff} = const.$ within the circulation. Thus the determination of the inhomogenity depending on time can be calculated just by one impulse distribution and its envelope curves. The intersection of the upper envelope curve (Fig. 43) which we use according to (82) with the line for $h \equiv \hat{h} = const.$ indicates on the abscissa the value of the required mixing time t_M. The relative mixing time $\tau_M = \frac{t_M}{\bar{t}_U}$ (88) is identical with the number i_M of circulations analogous to (78), required to give the prescribed degree of mixing and this in fact for geometric similar reactors independent of their size!

5.3 Residence-Time Behavior

When in continuous operation a through flow is superimposed on the circulation flow of the closed system (batch operation), then impulse responses as shown in Figs. 49 and 50 result. But also in this case the same relative mixing times τ_M are needed for a certain specified inhomogenity h, because, due to dilution by the unmarked throughflow, c drops in time t to the same extent as the reference variable c_∞ (77), hence at all times t the value of $c_r \equiv \frac{c}{c_\infty}$ (76) is the same as in batch operation.

Since after each circulation, tracer mass – injected at $t = 0$ according to an impulse function – divides in the same ratio as mass flows, only $\dot{M}_2 = (n_U - 1)\,\dot{M}_1$ (4) of the total flow $\dot{M}_3 = n_U\,\dot{M}_1$ (4) remains in the LR after each circulation (see Fig. 6). Thus the amount of tracer within the reactor decreases after each circulation in the ratio

$$\frac{\dot{M}_2}{\dot{M}_3} = \frac{n_U - 1}{n_U} \tag{89}$$

and thus after x_U completed circulations by $\left(\frac{n_U - 1}{n_U}\right)^{x_U}$.

Hence it follows that for continuous operation (81) has only to be multiplied by this value to give[31]:

$$c_r = \Sigma\,(c_r)_{x_U} = \sum_{x_U = \tau_U - 2}^{x_U = \tau_U + 2} \left(\frac{n_U - 1}{n_U}\right)^{x_U} \left(\frac{Bo}{4\,\pi\,\tau_U}\right)^{1/2} exp\left[-\frac{(x_U - \tau_U)^2}{4\,\tau_U}\,Bo\right]. \tag{90}$$

Replacing τ_U (78) by the relative residence time of the reactor

$$\tau \equiv \frac{t}{\bar{t}} = \frac{\tau_U}{n_U} \tag{91}$$

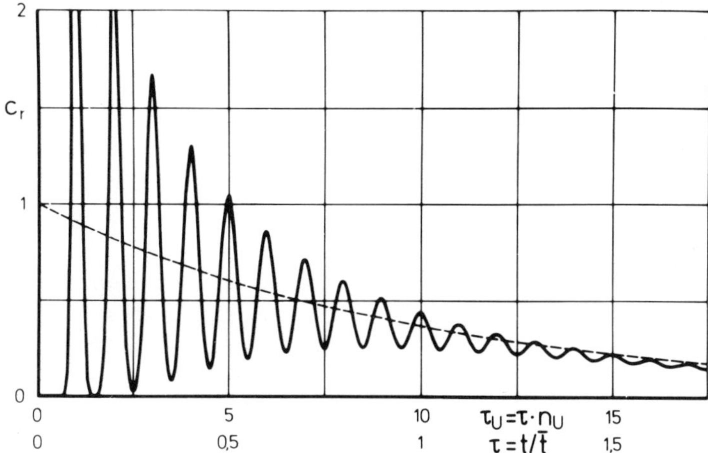

Fig. 49. Impulse response or distribution function of an *LR* with continuous operation (*Bo* = 200; $n_U = 10$)

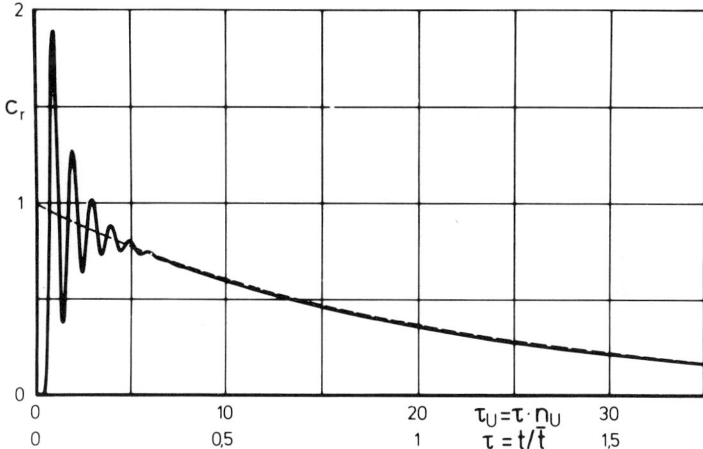

Fig. 50. Impulse response or distribution function of an *LR* with continuous operation (*Bo* = 50; $n_U = 20$)

Eq. (90) is identical to the *"distribution function"*

$$H(\tau) \equiv \frac{c_\tau}{c_\infty} \tag{92}$$

which as an *impulse response* characterizes the residence-time behavior of a continuous flow system as demonstrated in Figs. 49 and 50 with the τ-scale on the abscissa. The close connection between mixing- and residence-time behavior of continuous through flow systems is clearly demonstrated therewith!

The other common residence-time function, the *"transition function"* $S(\tau)$ of *LR* shall only be briefly mentioned here. It is considered in detail in[11].

The transition function is generally defined as

$$S(\tau) \equiv \int_{\tau=0}^{\tau} H(\tau)\, d\tau = \frac{M'}{M_0} \tag{93}$$

where

M': amount of tracer mass which has left the system up to time t

M_0: amount of tracer mass injected as an impulse at $t = 0$.

Thus it arises directly from the impulse response (90) and (92).

But $S(\tau)$ can also be regarded as a step-change response and be defined as

$$S(\tau) \equiv \frac{c_\tau}{c_{in}} \tag{94}$$

where

c_τ : tracer concentration in the outlet flow (through flow) at t

c_{in}: tracer concentration in the inlet flow (through flow) beginning stepwise at $t = 0$ and then remaining constant.

Notice the very different way of constant tracer input beginning at $t = 0$ and then remaining constant in this consideration compared with the formerly assumed impulse input only just at $t = 0$!

A mass balance of the tracer in the mass flows splitting after each circulation as shown in Fig. 6 for plug flow (i.e., no longitudinal mixing in the circulation, only backmixing due to recirculation) results in[11]

$$S(\tau_i) = 1 - \left(\frac{n_U - 1}{n_U}\right)^{n_U \tau_i} \tag{95}$$

with

$$\tau_i \equiv \frac{t_i}{\bar{t}} \tag{96}$$

and

$$t_i = i\, \bar{t}_U \tag{97}$$

where i is the number of *completed* circulations and t_i the time needed therefore with plug flow velocity $w \equiv w_m$ corresponding to (78).

The transition function $S(\tau_i)$ calculated as above for the *LR* with plug flow is plotted in Fig. 51 with the rotation number n_U as the decisive parameter. According to the derivation, it is only valid for the upper corners of the curve steps.

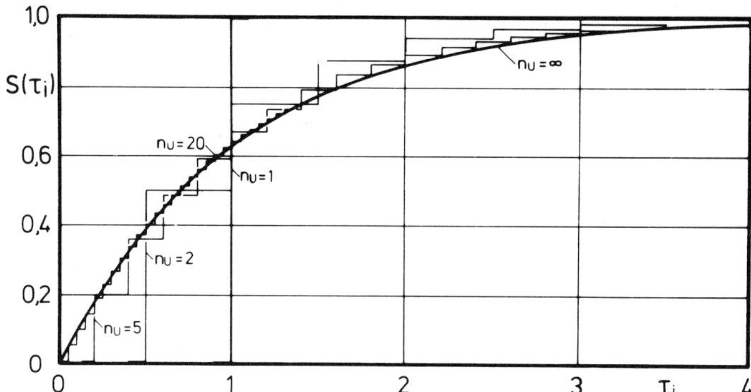

Fig. 51. Transition functions calculated for plug flow in *LR* with various n_U-values

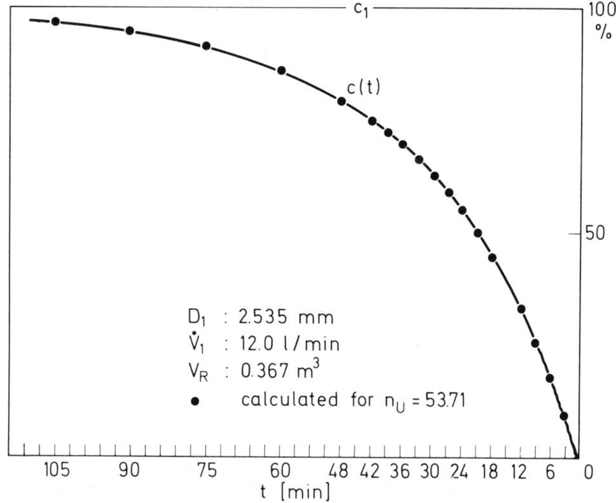

Fig. 52. Transition function measured in real *LR* (curve) with calculated points for $n_U = 53.71$ (95)

It is evident that

- $n_U = 1$ corresponds to an ideal tube (plug flow)
- $n_U \to \infty$ corresponds to an ideal tank (perfect mixing)
- $n_U \gtrsim 20$ gives reasonable approximation to the transition function of an ideal stirred tank, even without any longitudinal mixing in the circulation – just due to backmixing by recirculation.

The effect of additional longitudinal mixing improves this approximation for the real circulation flow in *LR*, as the measured transition function at $n_U = 53.71$ in Fig. 52 shows. The calculated points confirm good agreement between theory and experiment.

The residence-time behavior of chemical (and biochemical) reactors is extremely important for flow processes, since the conversion rate of (bio-)chemical reactions is time-dependent. Thus the molar concentration $c_1 = \frac{N_1}{\Delta V}$ of reactant 1 in the volume element ΔV of the throughput changes during its *individual* residence-time t in the reaction space for a first order reaction rate constant k' at $V_R = const.$ from the initial concentration c_{1_0} to the exit concentration

$$c_{1_t} = c_{1_0}\, e^{-k' t} . \tag{98}$$

This corresponds to a conversion rate of reactant A_1, in time t of

$$u_{1_t} = \frac{c_{1_0} - c_{1_t}}{c_{1_0}} = 1 - \frac{c_{1_t}}{c_{1_0}} = 1 - e^{-k' t} . \tag{99}$$

Now every continuous flow apparatus has indeed a mean residence-time defined by its mass content M_R and mass throughput $\dot{M}_1$ (8). However the various elements of the throughput have their own individual residence-time t which determines their individual conversion rate u_{1_t} differing more or less from the mean one. The mean total conversion rate $\overline{u}_1$ for a reactor with $\overline{t}$ is thus given by the summation (or integration) of the different conversions of all the elements by means of the *"conversion integral"*, which combines kinetics and residence-time behavior for first order homogeneous liquid phase reactions[33]:

$$1 - \overline{u}_1 = \int_{t=0}^{\infty} \frac{c_{1_t}}{c_{1_0}} H(t)\, dt . \tag{100}$$

6 Gas-Liquid-Interfacial Area (G-L-Interface)

6.1 Oxygen Transfer in Aerobic Bioreactions

In heterogeneous reaction systems in addition to the distribution of all reaction components within the reactor, which is characterized by flow, mixing, and residence-time behavior, the degree of dispersion of the dispersed phases is extremely important, since it determines the interfaces and thus the mass transfer between the phases.

The combined effects of dispersion and mixing are clearly illustrated by the example of the O_2-supply to the microorganisms in the *SCP*-process[1, 2]. For example, let the required specific oxygen demand referred to V_L be

$$\dot{m}_{O_2} = \frac{\dot{M}_{O_2}}{V_L} = 8 \text{ kg m}^{-3}\text{ h}^{-1} = 2.2 \text{ g m}^{-3}\text{ s}^{-1} . \tag{101}$$

Table 3

	Sulfite system (25 °C)	SCP system (37 °C)
He	67.2	40
D_L	1.86×10^{-9} m²/s	1.9×10^{-9} m²/s
k_L [a]	1.2×10^{-4} m/s	1.1×10^{-4} m/s
k_L [b]	4.6×10^{-4} m/s	

[a] Measured in falling film column (laminar annular film flow).
[b] Measured in loop reactor (turbulent bubble flow).

Even if the culture broth were saturated with O_2, i.e., sparging with air at p = 1 bar, T= 37 °C, and He = 40 according to Table 3[20])

$$\hat{c}_L = \frac{\hat{M}_{O_2}}{V_L} \approx 6.5 \text{ g m}^{-3} \tag{102}$$

it would only satisfy the O_2-demand for the short time of

$$t = \frac{\hat{c}_L}{\dot{m}_{O_2}} \approx 3 \text{ s} \tag{103}$$

after interruption of the O_2-supply.

On the other hand, the O_2-concentration c_L in the liquid (see Fig. 53) should be as low as possible, in order to maintain as high a mean concentration difference $\overline{\Delta c} = \overline{c}_o - c_L$ (108) as possible in the liquid film all over the G–L-interface.

Suppose the O_2-concentration in the culture broth be close to the vital minimum value

$$\check{c}_L = \frac{\check{M}_{O_2}}{V_L} \approx 1 \text{ g m}^{-3} \; (\hat{=} \; 1 \text{ ppm}) \, . \tag{104}$$

Then first of all the liquid phase must be intensely mixed throughout the whole reactor to avoid local concentration depressions below this minimum value.

Furthermore in any case the O_2-transfer from the air to the microorganisms adequate to their consumption must be maintained always and throughout the whole reaction space, because there is no considerable reserve in the liquid.

Based on the *"film theory"* (Fig. 53) the steps in this process are[2]):

- diffusion of O_2 through the G-film at the G–L-interface, which can be neglected here in consequence of the turbulence in the gas phase; therefore

$$c_{G_i} = c_G \tag{105}$$

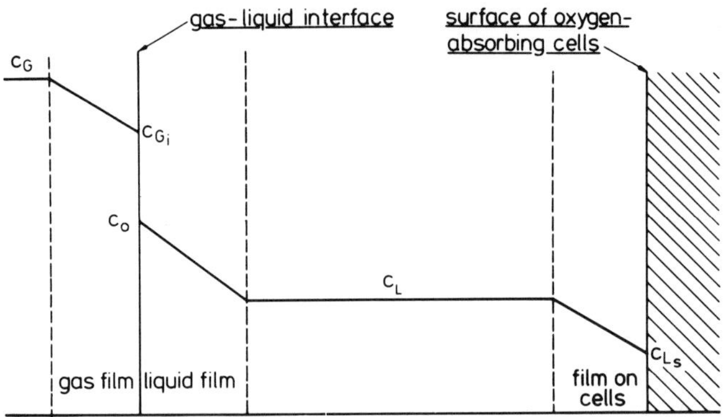

Fig. 53. Oxygen transfer from air to cell surface according to film theory

- absorption of O_2 across the G–L-interface A, which can be described, assuming constant equilibrium state according to Henry's law

$$c_o = \frac{c_{G_i}}{He} = \frac{c_G}{He} \tag{106}$$

- diffusion of O_2 through the L-film at the G–L-interface A. According to Fick's first law of diffusion the mass flow is

$$\dot{M}_{O_2} = - D_L \, A \left(\frac{dc}{dx}\right)_i . \tag{107}$$

For a linear gradient in the liquid film of thickness δ, as can be assumed for pure physical diffusion and approximately also for combined diffusion and slow chemical reaction, it follows with the mean concentration $\overline{c}_o$ along the whole G–L-interface a mean driving concentration difference

$$\overline{\Delta c} = \overline{c}_o - c_L \tag{108}$$

and a mass flow of O_2 according to (107) of

$$\dot{M}_{O_2} = \frac{D_L}{\delta} A \, \overline{\Delta c} . \tag{109}$$

With the mass transfer coefficient

$$k_L \equiv \frac{D_L}{\delta} \tag{110}$$

the specific O_2-input $\dot{m}_{O_2}$ (101) and the specific $G-L$-interface

$$a_L \equiv \frac{A}{V_L} \tag{111}$$

it follows from (109) that

$$\dot{m}_{O_2} = k_L\, a_L\, \overline{\Delta c} \tag{112}$$

- convective O_2-transport in the turbulent liquid phase of aequeous systems is so intense that c_L can be considered uniform there;
- diffusion of O_2 through the L-film at the cells to the $L-S$-interface, principally takes place as in the L-film at the $G-L$-interface, but can nevertheless be ignored according to the present state of knowledge, because the specific $L-S$-interface is about 10^6 times larger than the specific $G-L$-interface.
  Under these conditions, which are generally assumed at present, the overall O_2-transport from the air bubbles to the cell walls is practically controlled by the diffusion through the L-film at the $G-L$-interface and consequently determined by (112). The parameters involved in this equation are of decisive importance for selection, design and operation of the appropriate bioreactor. Therefore they have to be discussed a little more in this regard.
- $\overline{\Delta c}$ depends on the mixing in the phases and lies between the following limits:
  - completely segregated gas phase in which each gas bubble flows through the reactor without any interaction with others.
    It then follows[27]:

$$\overline{c}_o = \frac{c_{o_a} - c_{o_\omega}}{ln \dfrac{c_{o_a}}{c_{o_\omega}}} \quad {}^{a} \tag{113}$$

  - Completely mixed gas phase in which there is everywhere from inlet to outlet

$$\overline{c}_o \equiv c_{o_\omega} \tag{114}$$

Figure 54 shows that the difference between $\overline{c}_o$ (113) and c_{o_ω} (114) both referred to c_{o_a}, increases with increasing O_2-conversion rate which is defined corresponding to (99) as

$$\overline{u}_{O_2} \equiv \frac{c_{G_a} - c_{G_\omega}}{c_{G_a}} = 1 - \frac{c_{G_\omega}}{c_{G_a}}. \tag{115}$$

---

[a] indices a at air inlet
ω at air outlet

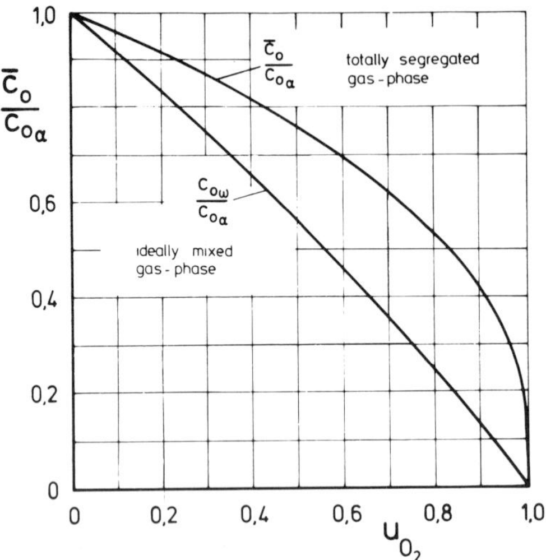

Fig. 54. Mean oxygen concentration $\overline{c}_O$ (113) and c_{O_ω} (114) referred to c_{O_a} as a function of oxygen conversion $\overline{u}_{O_2}$

For example at $\overline{u}_{O_2}$ = 50% one finds $\frac{\overline{c}_O}{c_{O_a}} \approx 0.76$ compared to $\frac{c_{O_\omega}}{c_{O_a}} \approx 0.57$[27]. The question of the really effective mean driving concentration difference $\overline{\Delta c}$ in *LR* still needs more detailed investigations of local concentration distributions. For the present we assume $\overline{c}_O$ (113) and therewith the largest possible $\overline{\Delta c}$ (108). Thus we determine with our chemical method (6.2) the smallest values of *G–L*-interface *A* (126), as explained there.
Assuming $c_L = \check{c}_L = 1 \text{ g m}^{-3}$, then at $\overline{u}_{O_2} \approx 0.5$ for the sulfite system approximately yields

$$\overline{\Delta c} = \overline{c}_O - c_L \approx 2.5 \times 10^{-3} \text{ kg m}^{-3} \tag{116}$$

- k_L (110) proves to be only slightly dependent on fluid dynamics *in highly turbulent systems.* This fact, which has been found by many observers, was also confirmed by our experiments in the sulfite system[27].
From these measurements with *JLR* we found a mean value according to Fig. 55 within the range investigated of

$$k_L \approx 4.6 \times 10^{-4} \text{ m s}^{-1} \approx 1.6 \text{ m h}^{-1} . \tag{117}$$

Own measurements in a falling film absorber at laminar flow gave

$$k_{L_l} \approx (1.1 - 1.2)\, 10^{-4} \text{ m s}^{-1} \tag{118}$$

for the *sulfite system* and nearly the same for an *SCP*-biosystem according to Table 3[19].

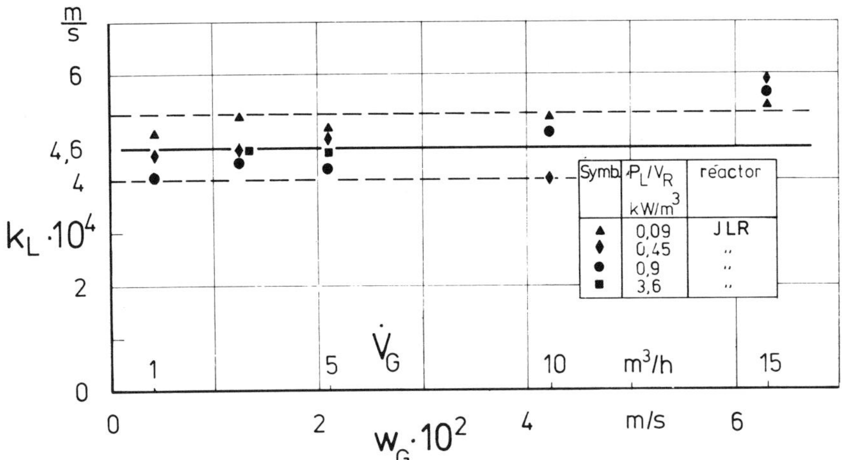

Fig. 55. Mass transfer coefficient k_L measured in *JLR* with sulfite system as a function of superficial gas velocity w_G

Since in the turbulent range of the *JLR* for both systems almost equal diffusion coefficients D_L according to Table 3, were measured, we assume k_L from (117) as an approximation for this *SCP*-biosystem too.

Considering for comparison Higbie's *"penetration model"*[34], modified by Danckwerts[35], with the *"mean equivalent diffusion time t_{eq}"* for instationary diffusion as proposed by Astarita[36], then the physical mass transfer coefficient is

$$k_{L_{Pen}} \equiv \left(\frac{D_L}{t_{eq}}\right)^{1/2} . \tag{119}$$

Without going into details, the characteristic dependence of $k_L \sim D_L$ (110) for the film model, and $k_{L_{Pen}} \sim D_L^{1/2}$ (119) for the penetration model shall be pointed out. According to Higbie's model one can consider t_{eq} as the mean residence-time of the liquid elements in the liquid boundary layer of a gas bubble rising with a mean velocity $\overline{w}_B$ and diameter $\overline{d}_B$ as

$$t_{eq} \approx \frac{\overline{d}_B}{\overline{w}_B} . \tag{120}$$

In the *sulfite system* values for mean bubble diameters of $\overline{d}_B \approx 2 - 4$ mm were determined. In the *JLR* at $\frac{P_L}{V_R} \approx 1$ kW m^{-3} and $\dot{V}_G = 3 - 10$ m^3 h^{-1}, for example, $\overline{d}_B \approx$ 2 mm[28] was found.

Also in the *SCP*-biosystem (methanol and ethanol basis)[37) the bubble diameter is close to these values. The rising velocity of the bubbles corresponding to this bubble diameter, $\bar{d}_B \approx 2$ mm is in the order of $\bar{w}_B \approx 200$ mm/s. This results in

$$t_{eq} \approx 10^{-2} \text{ s} .$$

The penetration model with $D_L \approx 2 \times 10^{-9}$ m s^{-1} (*sulfite system* at 30 °C, biosystem *SCP* at 37 °C, see Table 3) then leads by (119) to

$$k_{L_{Pen}} \approx \left(\frac{2 \times 10^{-9}}{10^{-2}}\right)^{1/2} \approx 4.5 \times 10^{-4} \text{ m s}^{-1} . \tag{121}$$

This rough estimate based on the *penetration model* and its comparison with experimental results based on the *film model* (117) demonstrates good agreement.

Similarly the estimate of the imaginary boundary layer thickness δ of the L-film in the *film model* from (110) and Table 3 may give an idea of the order of magnitude *for laminar systems*

$$\delta_l = \frac{D_L}{k_{L_l}} \approx 2.0 \times 10^{-5} \text{ m} = 20\ \mu\text{m} \tag{122}$$

and *for turbulent systems*

$$\delta_t = \frac{D_L}{k_{L_t}} \approx 4.4 \times 10^{-6} \text{ m} = 4.4\ \mu\text{m} . \tag{123}$$

Substituting $\dot{m}_{O_2} = 8$ kg · m^{-3} · h^{-1} (101); $\overline{\Delta c} = 2.5 \times 10^{-3}$ kg m^{-3} (116); $k_L = 1.6$ m h^{-1} (117) in (112) leads to the required specific G–L-interface in this example of

$$a_L = \frac{A}{V_L} = \frac{\dot{m}_{O_2}}{k_L\ \overline{\Delta c}} \approx 2000 \text{ m}^{-1} . \tag{124}$$

6.2 G–L-Interface in JLR and ALR

Thus the question arises: How can a_L be determined and how can it be definitely realized by the reactor design?

In our investigations the G–L-interface A was determined according to a combined experimental/theoretical method using the *sulfite system*[27, 28].

According to the film theory, the molar flow density $\dot{n}_{O_2}$ can be calculated in such a homogeneous liquid phase reaction by theoretically combining absorption of oxygen from air into liquid, its diffusion through the liquid film (see Fig. 53) combined with the liquid phase chemical reaction, resulting in

$$\dot{n}_{O_2} = \frac{\dot{N}_{O_2}}{A} = \bar{c}_O \left(\frac{2}{3} D_L\ k_2\ \bar{c}_O\right)^{1/2} (1 + C)^{1/2} . \tag{125}$$

All included values can be determined[27, 28]:

$\overline{c}_o$: mean concentration allover the G–L-interface according to (113) as explained there;

D_L: diffusion coefficient for O_2 in the *sulfite system* (Table 3);

k_2 : reaction rate constant of the sulfite oxidation, which is of 2nd order with respect to O_2[27, 28, 38–41];

C : integration constant[27].

On the other hand at steady state conditions, the actual molar flow $\dot{N}_{O_2}$ is measured in the experimental apparatus by a mass balance of the inlet and outlet air flows.

With the measured $\dot{N}_{O_2}$ and the calculated $\dot{n}_{O_2}$ (125) the interfacial area can be determined as

$$A \equiv \frac{\dot{N}_{O_2}}{\dot{n}_{O_2}} . \tag{126}$$

The combined experimental/theoretical determination of the G–L-interface A is always based here on $\overline{c}_o$ (113), which represents the largest possible value for given c_{o_α} and c_{o_ω}. Hence the largest possible value of the calculated $\dot{n}_{O_2}$ (125), and thus the smallest possible G–L-interface A (126) results from the measured $\dot{N}_{O_2}$. Therefore the values of the G–L-interface A determined with this chemical method are the smallest ones and therefore on the safe side. The more intense the mixing effect is, the larger the actual interface A_{real} is compared to A. Thus A_{real} will exceed A more in LR than for instance in bubble column reactors. That should always be considered in comparison of principally different reactor types.

Referring A to V_R or V_L leads to the *specific G–L-interface*

$$a_R \equiv \frac{A}{V_R} \tag{127}$$

or with (63) to

$$a_L \equiv \frac{A}{V_L} = \frac{a_R}{1 - \epsilon} . \tag{128}$$

Let us consider here a_R first. In Fig. 56 it is plotted against w_G with parameter $\frac{P_L}{V_R}$. Here also the lower curve (P_L = 0) is valid, as in Fig. 36 for the ALR (Re_1 = 0). It can be calculated in this range[28] by the linear relation

$$a_{R_{ALR}}\ [\mathrm{m}^{-1}] \equiv \frac{A}{V_R} \approx 4.3 \times 10^3\ w_G\ [\mathrm{m\ s}^{-1}] . \tag{129}$$

In the JLR ($P_L > 0$) up to $w_G \approx 1$ cm s^{-1} the gas flow is dispersed into very small bubbles by the liquid jet (steep rise of a_R). Then with increasing w_G, a_R increases ever

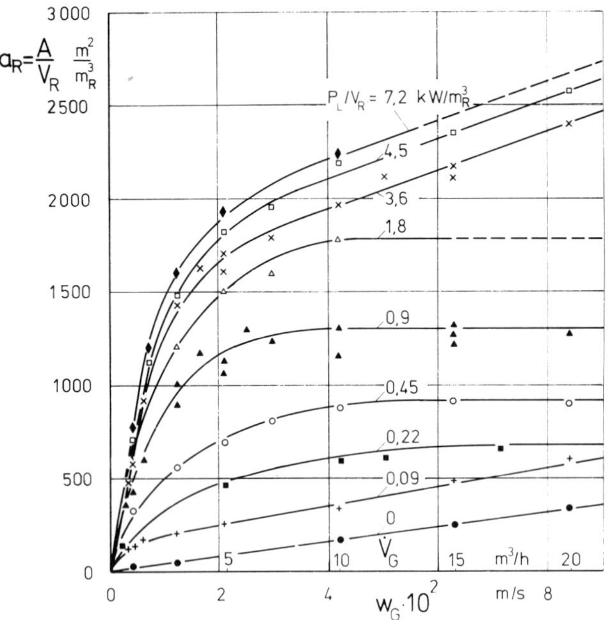

Fig. 56. Specific interfacial area $a_R = \dfrac{A}{V_R}$ in *LR* with *ring nozzle* sparger

slower and reaches a maximum value $\left(\text{at } \dfrac{P_L}{V_R} \lesssim 2 \text{ kW m}^{-3}\right)$, which can be considered to be a *"jet gas loading limit"* comparable to the *"system's gas loading limit"* in Fig. 36. When $\dfrac{P_L}{V_R} \gtrsim 2 \text{ kW m}^{-3}$, a_R shows a further flat linear increase.

The dependence of a_R on $\dfrac{P_L}{V_R}$ as was on the other hand ϵ on $\left(\dfrac{P_L}{V_R} D_1\right)^{1/3} \sim Re_1$ (Fig. 37) is evident from Fig. 57, where at $\dot{V}_G = 1; 3; 5 \text{ m}^3 \text{ h}^{-1}$ similar a_R-values were obtained for various nozzle diameters $D_1 = 1.85; 3.5; 7$ mm. For the range below the *"jet gas loading limit"* and for the linear increase of a_R in Fig. 57, the experimental results can be mathematically approached by

$$a_{R_{JLR}} \,[\text{m}^{-1}] \equiv \frac{A}{V_R} \approx 5.4 \times 10^3 \, w_G^{0.4} \left(\frac{P_L}{V_R}\right)^{0.66} \tag{130}$$

where

$$a_{R_{JLR}} \,[\text{m}^{-1}];\; w_G \,[\text{m s}^{-1}];\; \left(\frac{P_L}{V_R}\right) [\text{kW m}^{-3}]\,.$$

The excellent dispersion effect of the highly turbulent liquid jet is demonstrated in Fig. 58. In *ALR* operation $\left(\dfrac{P_L}{V_R} = 0\right)$, a_R increases rather slowly with $\dfrac{P}{V_R} = \dfrac{P_L + P_G}{V_R}$ for all gas

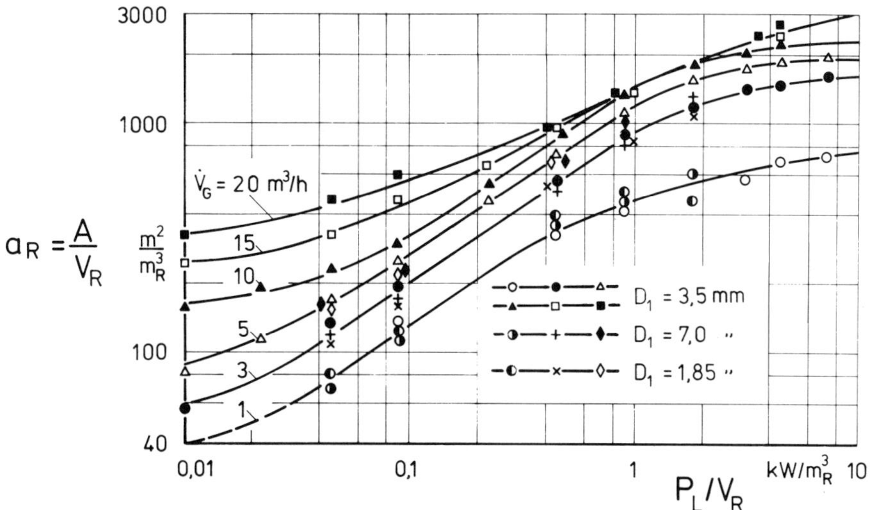

Fig. 57. Specific interfacial area a_R in *JLR* as a function of specific liquid power input $\frac{P_L}{V_R}$

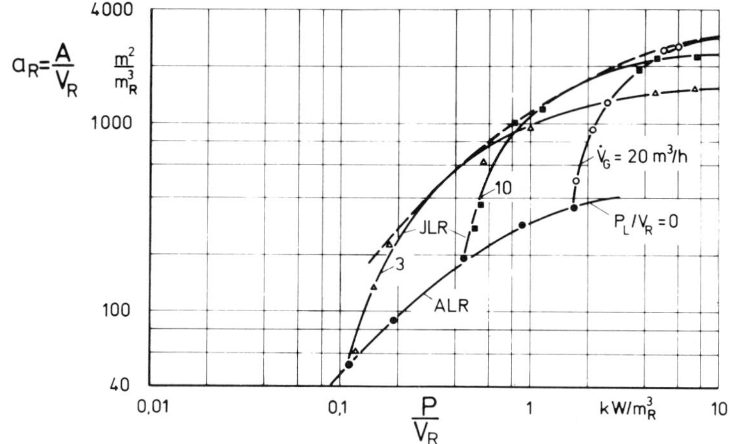

Fig. 58. Specific interfacial area a_R in *ALR* and *JLR* as a function of the total specific power input $\frac{P}{V_R}$

flows $\dot{V}_G$. If one now changes from *ALR* to *JLR* operation, for example at $\dot{V}_G = 10\ m^3\ h^{-1} \triangleq w_G = 4.2\ cm\ s^{-1}$ then a_R increases very steeply and reaches much higher values, e.g., $a_R \approx 1000\ m^{-1}$ when $\frac{P}{V_R} \approx 1\ kW\ m^{-3}$, whereas with *ALR* at the same power input only $a_R \approx 300\ m^{-1}$ was reached in the same *sulfite system.*

This is valid in the same way for a_L. From Fig. 59 one can see that the required value of $a_L \approx 2000\ m^{-1}$ which was roughly estimated in (124) can be reached with the

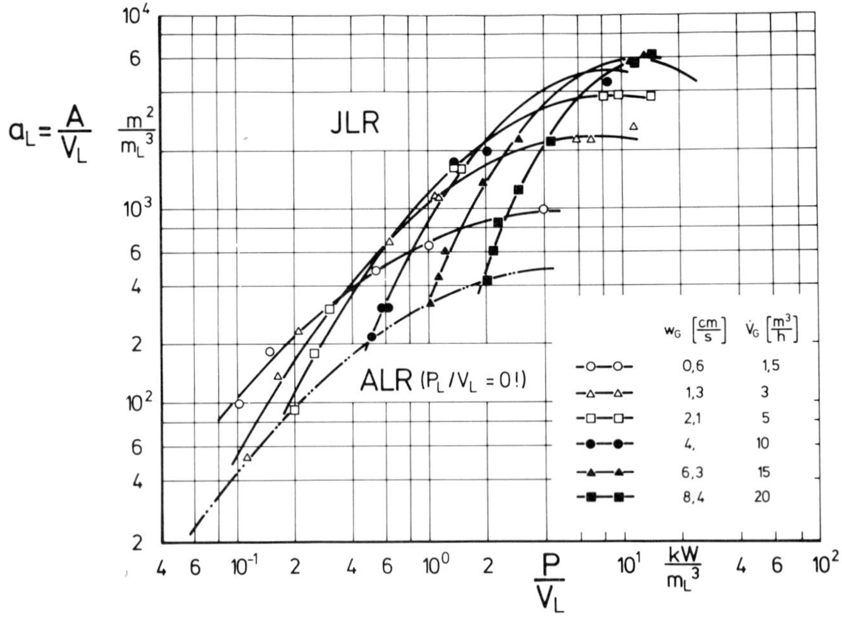

Fig. 59. Specific interfacial area a_L in *ALR* and *JLR* as a function of total power input $\frac{P}{V_L}$

JLR in the *sulfite system* by a specific total power input of $\frac{P}{V_L} \approx 2$ kW m^{-3}. Correspondingly (128) would give, e.g., at a gas content of $\epsilon = 0.5$, a required specific *G–L*-interface referred to V_R of $a_R \approx 1000$ m^{-1} which – as shown already in Fig. 58 – could be achieved with $\frac{P}{V_R} \approx 1$ kW m$^{-3} \triangleq \frac{P}{V_L} \approx 2$ kW m^{-3}.

Results of pre-investigations with *ALR* and *sulfite system* at high gas flow rates are presented in Fig. 60 (see 3.4.2). Sparging with an open tube at the bottom of the *ALR* ($H_B = 0$) results here at $\frac{P}{V_L} \triangleq \frac{P_G}{V_L} = 1$ kW m^{-3} in approximately the same specific *G–L*-interface $a_L \approx 350$ m^{-1} at $w_G \approx 8$ cm s^{-1} as in Fig. 59 at $w_G \approx 6.3$ cm s^{-1}. The specific *G–L*-interface of *ALR* can indeed be further increased above the low a_L-values of Fig. 59 by increasing w_G (Fig. 60), however the specific power demand ($P \equiv P_G$) also increases considerably. For example $a_L = 1000$ m^{-1} requires $\frac{P}{V_L} \approx 4$ kW m^{-3} at $w_G \approx 30$ cm s^{-1}. The value of $a_L = 2000$ m^{-1} determined in (124) could not be realized with *ALR* in our experiments (Fig. 60). By extrapolation of the measured curves approximately $\frac{P}{V_L} \approx 10$ kW m^{-3} and high superficial gas velocities of $w_G \gtrsim 60$ cm s^{-1} may be required therefore.

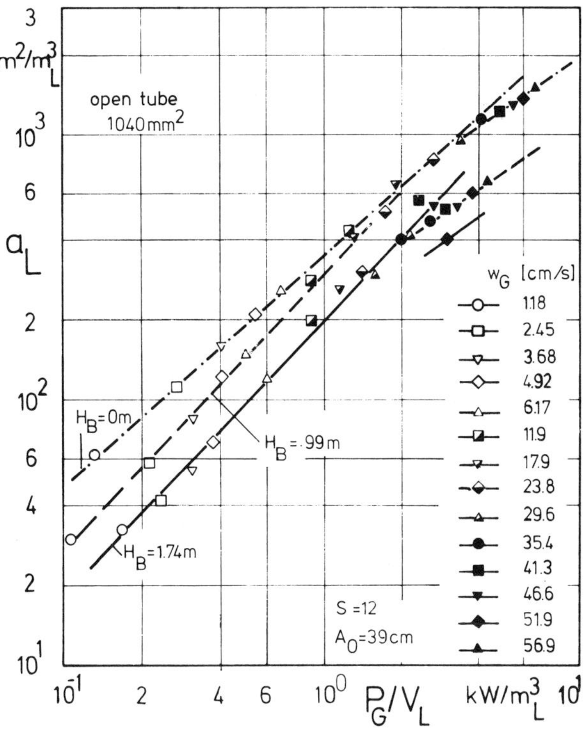

Fig. 60. Influence of superficial gas velocity w_G and aeration height H_B on interfacial area $a_L = \frac{A}{V_L}$ in *ALR*

Table 4 includes the interfacial areas per unit of power input for several important reactor types and comparable heterogeneous *G–L*-systems. This relation has the same value whether A as well as P are referred to V_R or V_L [28, 42].

These $\frac{A}{P}$-values are well confirmed for *JLR* by the upper envelope curve in Fig. 59 in the range $\frac{P}{V_L} \approx 0.1$–$4$ kW m^{-3} and for *ALR* by the lower dashed curve over the whole measured range. Figure 60 also confirms these values of $\frac{A}{P}$ for *ALR* in the much larger range up to $w_G \approx 60$ cm s^{-1} and $\frac{P}{V_L} \approx 10$ kW m^{-3}; and Fig. 61 shows this explicitly. From Figs. 60 and 61 one sees that with *ALR* one can increase a_L in *sulfite systems* possibly up to some 10^3 m^{-1} by increasing approximately w_G above 50 cm s^{-1} and $\frac{P}{V_L}$ above 10 kW m^{-3}.

Assuming the same value of $k_L = 1.6$ m h^{-1} (117) for *ALR* and *JLR* with highly turbulent flow, the following values of $k_L\, a_L$ were obtained in our experiments with the

Table 4

Gas liquid reactor type	Interfacial area per unit of power input $\frac{A}{P}\left[\frac{m^2}{kW}\right]$	Oxygen transfer rate per unit of power input $\frac{\dot{M}_{O_2}}{P}\left[\frac{kg}{kWh}\right]$
Mammoth-loop reactor MLR	300– 500	1,2–2,0
Bubble column reactor BCR	200– 600	0,8–2,4
Stirred tank reactor STR	300– 600	1,2–2,4
Jet-loop reactor JLR	1,000–1,200	4,0–6,0

sulfite system at a specific total power input of $\frac{P}{V_L} = 2\ \text{kW m}^{-3}$:

$$\begin{aligned} &ALR \text{ roughly} \quad k_L\, a_L \approx 1000\ \text{h}^{-1} \approx 0.27\ \text{s}^{-1} \\ &JLR \text{ roughly} \quad k_L\, a_L \approx 3600\ \text{h}^{-1} = 1\ \ \text{s}^{-1}\,. \end{aligned} \tag{131}$$

Assuming $k_L = 1.6\ \text{m h}^{-1}$ (117) and $\overline{\Delta c} = 2.5 \times 10^{-3}\ \text{kg m}^{-3}$ (116) for all types of reactors considered, the specific rate of O_2-input is according to (112)

$$\dot{m}_{O_2}\ [\text{kg m}^{-3}\ \text{h}^{-1}] = k_L\ \overline{\Delta c}\ a_L = 4 \times 10^{-3}\ a_L\,. \tag{132}$$

Thus in the example considered above $\dot{m}_{O_2} = 4 \times 10^{-3} \times 2000 = 8\ \text{kg m}^{-3}\ \text{h}^{-1}$ (101). The specific rate of O_2-input $\dot{m}_{O_2} = \frac{\dot{M}_{O_2}}{V_L}$ referred to the specific total power input $\frac{P}{V_L}$ [kW m^{-3}] is then

$$\frac{\dot{M}_{O_2}}{P}\ [\text{kg kW}^{-1}\ \text{h}^{-1}] = k_L\ \overline{\Delta c}\ \frac{A}{P} = 4 \times 10^{-3}\ \frac{A}{P}\,. \tag{133}$$

Hence with $\frac{A}{P} = 1000\ \text{m}^2\ \text{kW}^{-1}$ (Table 4) arises $\frac{\dot{M}_{O_2}}{P} = 4\ \text{kg kW}^{-1}\ \text{h}^{-1}$. These values are also included in Table 4.

For the *JLR* and *sulfite system* they are in the order of magnitude

$$\frac{\dot{M}_{O_2}}{P} \approx 4 - 6\ [\text{kg kW}^{-1}\ \text{h}^{-1}]\,. \tag{134}$$

One can qualitatively conclude that for gas flow rates with $w_G \leqslant 10\ \text{cm s}^{-1}$ the *JLR* produces about 3–4 times greater specific *G–L*-interfaces than the *ALR* at the same total power input (Fig. 59). When with increasing gas flow in *JLR* the *"jet gas loading limit"* is reached, the gas flow should be split up to several liquid jets[12].

The *ALR* appears to be especially suitable for high gas flow rates, however it then requires higher power input per unit of *G–L*-interface $\frac{P}{A}$ (Fig. 61).

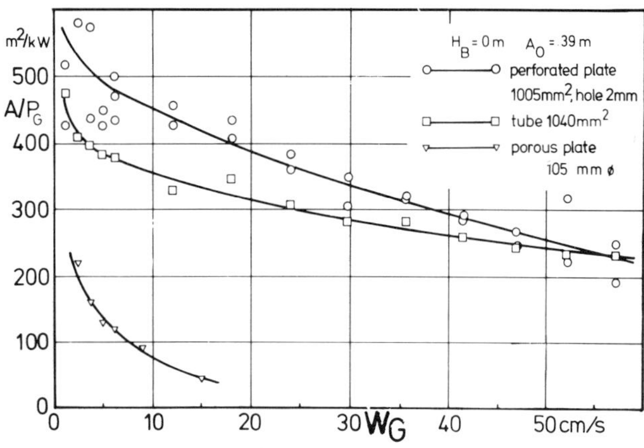

Fig. 61. Influence of w_G and the type of gas sparger on $\frac{A}{P_G}$ in ALR

6.3 JLR with Reversed Flow Direction

Other pre-investigations[19] were carried out concerning fluid dynamics with reversed flow direction as shown in Fig. 62. In this case liquid jet and gas were not introduced in the lower section in a vertical direction *upwards*, but as high as possible with a *downwards* directed liquid jet and gas input. Here as little as possible of the gas rising in the annulus should be re-circulated into the internal space of the draft tube, so that, by maintaining a large density difference in the two communicating spaces mainly above the gas input, as strong an air-lift drive as possible is effected in addition to the jet drive. This mode of operation has mainly the following advantages:

- the gas must flow as long a distance as possible within the draft tube and can only leave the reactor after rising through the annulus. Compared to the case of flow in the other direction, in which a part of the gas introduced at the bottom can already leave the reactor after rising through the draft tube and only a partial flow is re-circulated through the annulus, the residence-time of the gas can be increased and uniformed and hence (113) is approached.
- the hydrostatic pressure at this gas inlet is lower than at the bottom (Fig. 62), which means that the compression power for the gas flow input is lower. However, it should be pointed out that this spared air compression must be produced by the circulation energy of the liquid flow in the downwards directed flow and thus reduces the circulation intensity.

The experiments were carried out with water-air-system in the *JLR* 630 (Fig. 62) with constant values of D = 630 mm; D_1 = 10 mm; A_u = 21.5 cm; A_o = 69.5 cm; L_E = 1.81 m; H = 2.72 m; $s = \frac{H}{D}$ = 4.3; V_R = 0.848 m^3. The following parameters were varied: $\frac{D_E}{D}$ = 0.57 and 0.31; A_D = 0–1500 mm; $\frac{P_L}{V_R}$ = 0.07–1.5 kW m^{-3} and the gas sparger as shown in Fig. 63.

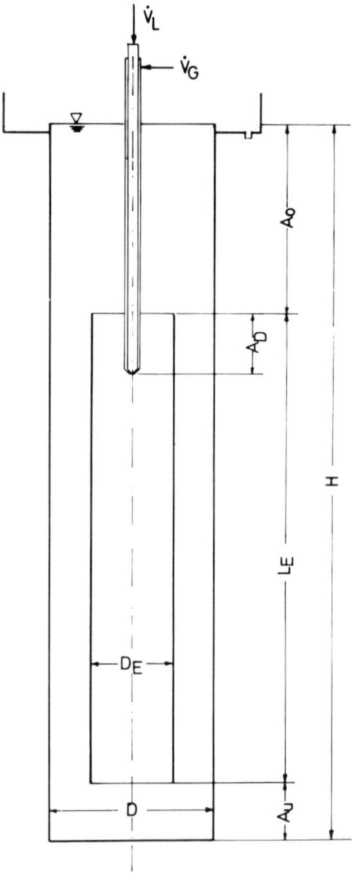

Fig. 62. Scheme of model reactor *JLR 630* for reversed flow direction

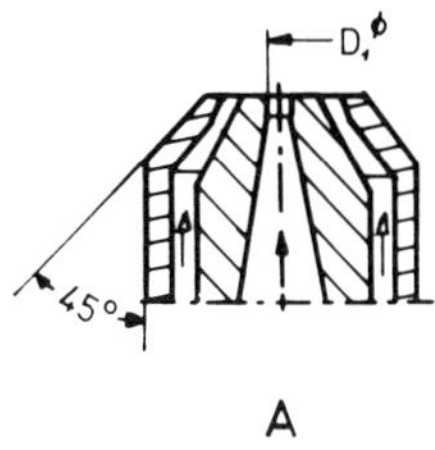

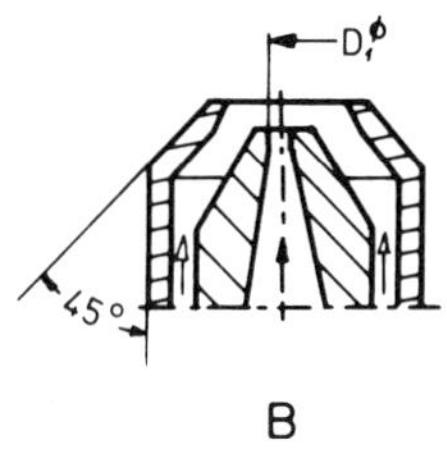

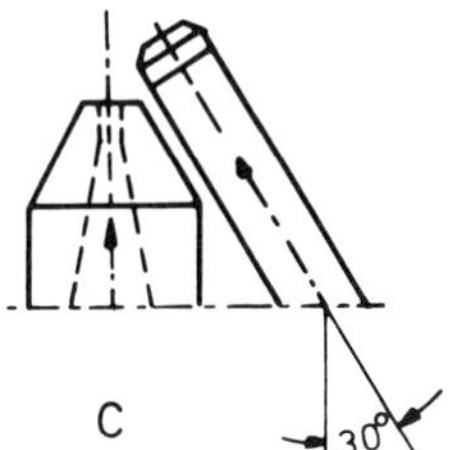

Fig. 63. Gas distributors used in the *JLR 630* (Fig. 62)

At the beginning of an experiment only the liquid jet was injected until a constant liquid circulation was established. Then air was introduced and its volume rate increased stepwise by $\Delta \dot{V}_G = 0.25\ m^3\ h^{-1}$.

At a maximal gas flow rate $\hat{\dot{V}}_G$, which can be exactly determined, the gas bubbles are no longer sucked downwards and led round the lower edge of the draft tube by the cir-

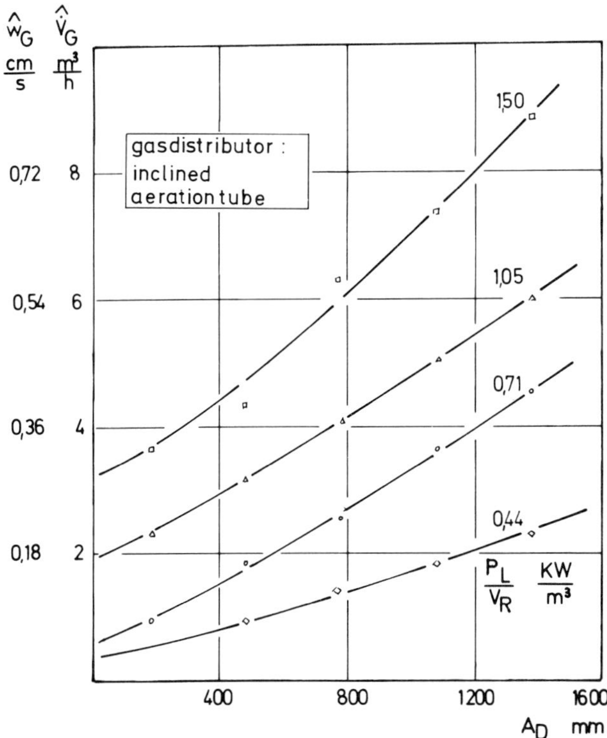

Fig. 64. Maximum gas flow $\hat{\dot{V}}_G$ as a function of jet nozzle distance A_D (Fig. 62) and specific power input $\frac{P_L}{V_R}$ $\left(\frac{D_E}{D} = 0{,}57\right)$

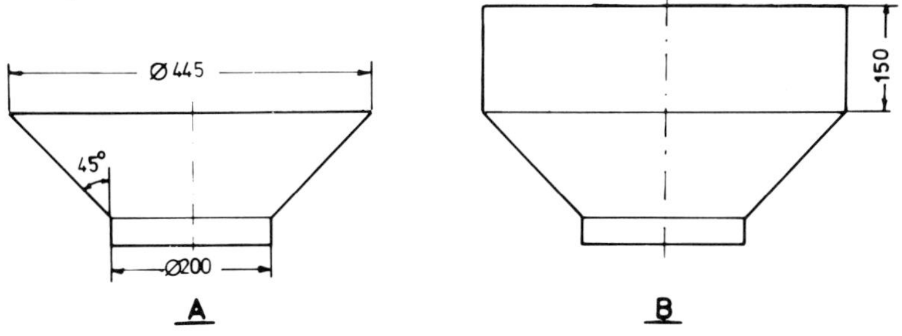

Fig. 65. Schematic figures of gas deflectors

culation flow, but rise upwards within the draft tube. Thus the direction of the air-lift drive reverses and opposes that of the liquid jet. As a result no definite liquid circulation occurs anymore.

In all cases it was found that sparging with the inclined aeration tube, as shown in Fig. 63, was the best one, and we will therefore limit our discussion to this type. Figure 64 shows that $\hat{\dot{V}}_G$ increases almost linearly with increasing A_D (Fig. 62). But the specific

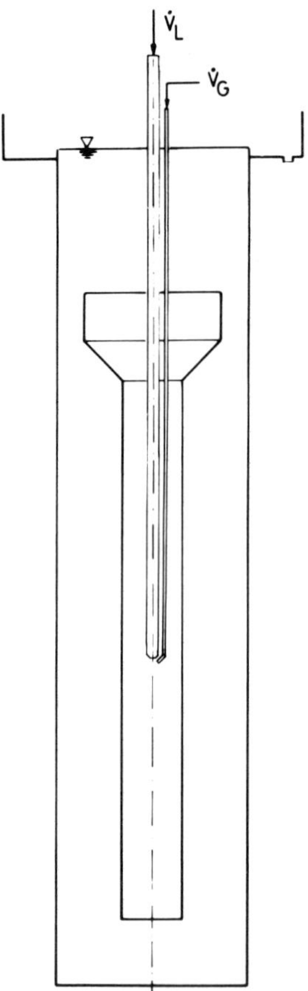

Fig. 66. Schematic figure of model reactor *JLR 630* with gas deflector for reversed flow direction

power demand $\frac{P}{V_R}$ is very high when referred to the small flow rates $\hat{\dot{V}}_G$ ($\hat{w}_G$) of entrained gas. This is due to the fact that on the one hand in the relatively large draft tube at $\frac{D_E}{D} = 0.57$ only a small downwards directed flow velocity was reached to draw the gas downwards; and on the other hand also a strong recirculation of the gas rising in the annulus back into the draft tube lowered the air-lift drive. Thus considerable improvements could be achieved by reducing $\frac{D_E}{D}$ to 0.31 and installing gas deflectors on the upper edge of the draft tube as shown in Figs. 65 and 66. This is clearly shown in Fig. 67. The effect of reducing D_E without gas deflectors is presented by the lower curves. It led to a reduction of the required specific power input $\frac{P}{V_R}$ of about 50% to achieve the same $\hat{\dot{V}}_G$.

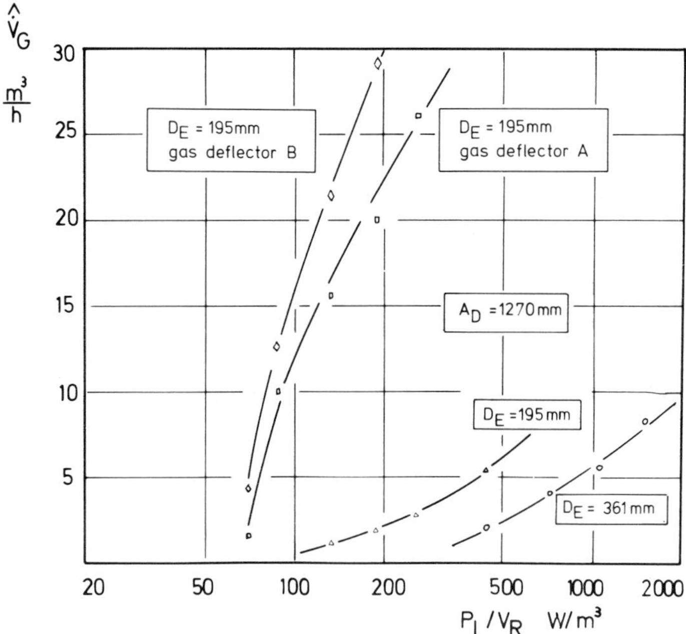

Fig. 67. Maximum gas flow $\hat{\dot{V}}_G$ as a function of power input (gas distributor *C* from Fig. 67)

In comparison, the upper curves demonstrate the extraordinary effect of gas deflectors on $\hat{\dot{V}}_G$ ($\hat{w}_G$), attained by the same power input $\frac{P_L}{V_R}$.

In this case the density difference between the *G–L*-mixture in the annulus and the liquid, nearly free of gas, inside the draft tube above the gas inlet together with the lifting force of the gas bubbles in the annulus cause an intense air-lift drive, which becomes the stronger the more the distance A_D between gas inlet and the upper edge of the draft tube is increased. Although the air-lift drive thus considerably contributes to the circulation flow, the liquid jet is of essential importance partially with its additional driving force, but even more because of its dispersion effect. Without the latter effect the circulation could only take up considerably less gas, because the entering large gas clusters would rise against the liquid down flow, lowering and finally stopping it. This was clearly demonstrated by switching off the liquid jet during sparged circulation.

This mode of operation of the *JLR* opens up interesting possibilities of realizing a *liquid circulation flow (LR-type)* and a *gas through flow (tube type)*; however according to our experiences it is in this way confined to small gas flow rates (see another possibility in Sect. 9.2).

7 Heat Transfer, Limiting Capacity, and Stability Behavior of LR

In (bio-)chemical – as in nuclear reactors –, the mass conversion is always accompanied by energy conversion. If we consider here – as mentioned in Sect. 1 – exothermic reactions, then in addition to the reaction enthalpy, also the heat dissipated by the mixing and dispersion power input, thus a total specific heat generation $\dot{q}$ [kW m^{-3}] must be removed per unit of reaction space. The discussion will be limited here to contact heat transfer across cooling areas[7)].

Since for contact cooling the heat which can be transferred is proportional to the cooling area $A_C \sim D^2$ and that generated in the reactor proportional to the reaction volume $V_R \sim D^3$, the specific cooling area and hence the possible heat transfer decreases for geometric similar scale-up by

$$\frac{A_C}{V_R} \sim D^{-1} . \tag{135}$$

Thus a limiting reactor size and production capacity results, at which the heat generated can just be removed. How can this limiting size be calculated and influenced by geometrical and heat transfer parameters?

Let us consider the *PLR* shown in Fig. 5 with external cooling of the jacket (A_M) and bottom (A_B), as well as internal cooling by the double-walled draft tube (A_E). Then with

$$s = \frac{H}{D} ; d_E = \frac{D_E}{D} ; l_E = \frac{L_E}{D}$$

for the cooling areas and the reactor volume the following is valid with reasonable accuracy:

$$A_B \approx \frac{\pi}{4} D^2 \ ; \ A_M \approx \pi\, s\, D^2 \ ; \ A_E \approx 2\, \pi\, d_E\, l_E\, D^2 \ ; \ V_R \approx \frac{\pi}{4}\, s\, D^3 .$$

The specific cooling area of the reactor is thus

$$a_C \equiv \frac{\Sigma A i}{V_R} = \frac{\frac{\pi}{4} D^2 \, (1 + 4\, s + 8\, d_E\, l_E)}{\frac{\pi}{4}\, s\, D^3} = \frac{1 + 4\, s + 8\, d_E\, l_E}{s\, D} = \frac{B_g'}{D} \tag{136}$$

with

$$B_g' \equiv \frac{1 + 4\, s + 8\, d_E\, l_E}{s} . \tag{137}$$

According to *Fourier's law* the specific caloric power $\dot{q}_C$, which can be removed from the reaction space by contact cooling is

$$\dot{q}_C = k\, a_C\, \Delta T_m \tag{138}$$

where

k [kW m^{-2} K^{-1}] is the overall heat transfer coefficient

ΔT_m [K] is the logarithmic mean temperature difference

between the reaction system ($T_R \approx$ const.) and the mean value of the cooling medium, e.g., cooling water ($\overline{T}_C$) i.e.,

$$\Delta T_m = T_R - \overline{T}_C\,. \tag{139}$$

Substituting (136) in (138) one obtains with maximal values of k and ΔT_m the maximal possible reactor diameter, up to which the specific heat generation $\dot{q}$ can just be removed ($\dot{q}_C = \dot{q}$)

$$\hat{D} = B_g' \left(\frac{\widehat{k\,\Delta T_m}}{\dot{q}}\right)\,. \tag{140}$$

The corresponding maximal reactor volume is

$$\hat{V}_R = \frac{\pi}{4}\, s\, \hat{D}^3 = \frac{\pi}{4}\, s\, B_g'^3 \left(\frac{\widehat{k\,\Delta T_m}}{\dot{q}}\right)^3 = B_g \left(\frac{\widehat{k\,\Delta T_m}}{\dot{q}}\right)\,. \tag{141}$$

For the *PLR* of Fig. 5 considered here, the characteristic geometric number

$$B_g \equiv \frac{\pi}{4}\, s\, B_g'^3 = \frac{\pi}{4}\, \frac{(1 + 4\, s + 8\, d_E\, l_E)^3}{s^2} \tag{142}$$

quite clearly only depends on dimensionless geometric parameters, which are all referred to D.

For certain caloric parameters $\hat{k}$; $\widehat{\Delta T}_m$; $\dot{q}$, from (141) arises

$$\hat{V}_R \sim B_g\,. \tag{143}$$

Figure 68 shows the magnitude of B_g and its dependence on the reactor type and design and especially on the grades of slenderness $s = \frac{H}{D}$. For a conventional *STR* (I) with $s \approx 1$ and external cooling $B_g \approx 100$ yields, for an *LR* with additional internal cooling of the draft tube (II) at $s = 5$ however $B_g \approx 2700$. Thus according to (141) a 27 times larger reaction volume $\hat{V}_R$ can be realized with this slender *LR* than with the convention-

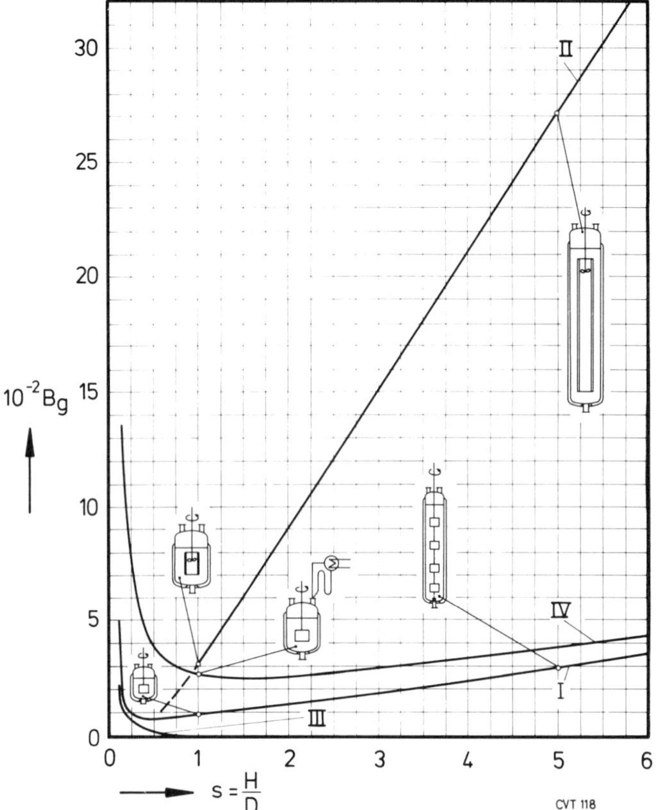

Fig. 68. Characteristic geometric number B_g for various reactor types and cooling systems depending on the grade of slenderness $s = \frac{H}{D}$

al *STR* at equal specific heat generation $\dot{q}$ and cooling conditions k and ΔT_m. This clearly demonstrates the advantage of the slim tower type *LR* with additional cooling of the draft tube concerning specific heat removal[7)]. This aspect can be important when very large reactors with high specific heat generation are desired without additional external cooling, as it is the case for instance with the *SCP*-reactor.

Example: According to our experiments[42)] the surface heat transfer coefficient within the *JLR* for a water-air-system at $Re_1 > 10^5$ is approximately $a_i \approx 3 - 4$ [kW m^{-2} K^{-1}]. Considering further heat transfer resistance (heat conduction in the wall, heat transfer to the cooling medium) let us assume an overall heat transfer coefficient of about $k \approx 2$ kW m^{-2} K^{-1}. Then, e.g., for the *SCP*-system at $\epsilon = 0.5$ may be: $\dot{q} = \dot{q}_C$ = 20 kW m^{-3} referred to V_R and ΔT_m = 8 K, hence $\left(\frac{k\,\Delta T_m}{\dot{q}}\right)^3 = 0.512$.

With B_g = 2700 for the *LR* at s = 5 according to (141) the maximal reactor volume, up to which $\dot{q}$ can be transferred ($\dot{q}_C = \dot{q}$), is $\hat{V}_R \approx 1382$ m^3.

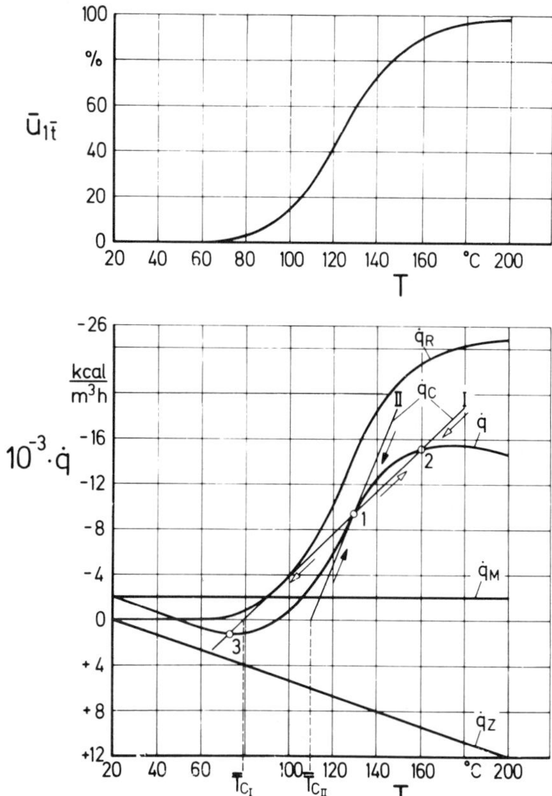

Fig. 69. Stability behaviour of chemical reactors with exotherm reaction

Although such a large *LR* would surely not be geometrically similar to that shown in Figs. 5 or 7, this rough estimate may at least indicate the order of magnitude for scale-up with respect to limiting heat transfer.

(141) is especially important too for the stability behaviour of the reactor with exothermic reactions[7)]. Figure 69 shows the total specific heat generation $\dot{q}$ at various reaction temperatures $T = T_R$ (abscissa). Two *"cooling lines"*, $(\dot{q}_C)_I$ and $(\dot{q}_C)_{II}$ which represent the specific heat removal $\dot{q}_C$ at T_R with $\overline{T}_{C_I} = 80$ °C respectively $\overline{T}_{C_{II}} = 110$ °C as abscissa values. For the possible operating points *1* and *2* at which $\dot{q} = \dot{q}_C$ yields according to (138) and (139)

$$\dot{q} = \dot{q}_C = k\, a_C\, \Delta T_m = k\, a_C\, (T_R - \overline{T}_C)\,. \tag{144}$$

The *"cooling lines"* are defined by two points:

- the optional abscissa values (here for example)

$$\overline{T}_{C_I} = 80\ °\text{C or } \overline{T}_{C_{II}} = 110\ °\text{C} \tag{145}$$

- the determined operating point *1*, here at T_R = 130 °C and

$$\dot{q} = \dot{q}_C = -9349 \text{ kcal m}^{-3} \text{ h}^{-1} = -10.87 \text{ kW m}^{-3} \tag{146}$$

where the negative sign indicates exothermic heat.

Let us consider first *cooling line I*. It has three stationary operating points *1, 2,* and *3,* at which $\dot{q}$ produced equals $\dot{q}_{C_I}$ removed. However, are they also *stable* in the sense of a self-stabilizing system?

If the temperature of the system is increased above T_R (determined operating point *1*) due to some disturbance, then more heat is produced ($\dot{q}$) than removed ($\dot{q}_{C_I}$). Thus the temperature increases further up to point *2*. If it exceeds this point, then cooling ($\dot{q}_{C_I}$) is larger than heat production ($\dot{q}$). Hence the system brings itself back to operating point *2*. Thus point *1* is unstable, but point *2* and in the same way point *3* are stable. The (static) stability criterion is apparently, that at the stationary operating point the cooling line ($\dot{q}_C$) must be steeper than the heat production curve ($\dot{q}$), i.e.,

$$\left(\frac{d\dot{q}_C}{dT}\right)_{T_R} > \left(\frac{d\dot{q}}{dT}\right)_{T_R} \tag{147}$$

or with (139)

$$\frac{\dot{q}_C}{\Delta T_m} > \left(\frac{d\dot{q}}{dT}\right)_{T_R} . \tag{148}$$

Now we tranform (141) with $\dot{q} = \dot{q}_C$ at the stationary operating point into

$$\hat{V}_R = B_g \frac{\hat{k}^3}{(\dot{q}_C/\Delta T_m)^3} . \tag{149}$$

Putting then the stability criterion (148) into (149) leads to the maximal possible reactor volume $\hat{V}_R$, which enables *stationary self-stabilizing* operation.

8 Economic Optimal Size of LR

The question now arises, whether the maximal possible $\hat{V}_R$ is also the most economic one. With a faster circulation flow – i.e., larger w_m and Re_m (11) – one can increase the surface heat transfer coefficient a_i within the reactor and therewith the overall heat transfer coefficient k too. However, that requires a larger specific power input $\frac{P}{V_R}$, which is then dissipated into heat and must be in addition removed as such. In consequence finally the total specific heat generation $\dot{q}$ exceeds the specific cooling capacity $\dot{q}_C$ of the

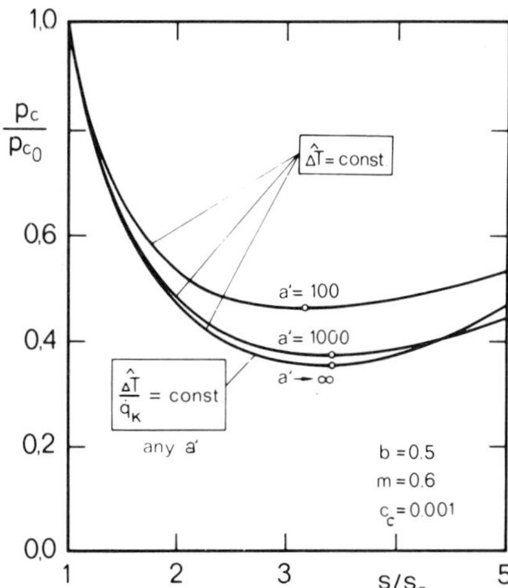

Fig. 70. Relative total costs $\frac{p_c}{p_{c_0}}$ of *JLR* as a function of the relative grade of slenderness $\frac{s}{s_0}$, both related to the *"reference JLR"* with $s_0 = 5$

reactor. Further increase of power input for faster circulation would then lead to a decrease of $\hat{V}_R$ (141) because $\dot{q}$ would increase more than k.

With increasing $\dot{V}_R$ the capital costs *per mass unit of product* only increase proportional to V_R^{m-1} with $m \approx 0.6$. However, the energy costs increase for the higher specific power input, which enables the growth of $\hat{V}_R$. An investigation of this optimization[44, 45], which can not be presented here in detail, shows the dependence of the relative total costs $\frac{p_c}{p_{c_0}}$ on the relative slenderness ratio $\frac{s}{s_0}$ of a *JLR* as shown in Fig. 70. Here p_{c_0} represents the total costs of the *"reference JLR"* with $s_0 = 5$. The parameter a' is defined as

$$a' \equiv \frac{\dot{q}_{RA}}{\dot{q}_{M_0}} \,. \tag{150}$$

$\dot{q}_{RA}$ combines the constant specific reaction heat generation $\dot{q}_R$ with $\dot{q}_A$ as the change of heat power of the entering mass flow in consequence of different temperatures of the inflow and the reaction system. $\dot{q}_{M_0}$ is the constant minimal mixing power – according to the demanded degree of mixing (Sect. 5.2) – of the *"reference JLR"* with $s_0 = 5$.

$b = 0.5$ means, that the heat transfer resistances within the wall and between wall and cooling medium are half of that at the internal surface.

$c_c = 0.001$ is the relation between energy and capital costs at the *"reference JLR"* with $s_0 = 5$.

Figure 70 clearly shows that with an increasing part of $\dot{q}_M$ or $\dot{q}_{M_0}$ at constant $\dot{q}_{RA}$ in the total specific heat generation

$$\dot{q} = \dot{q}_{RA} + \dot{q}_M \tag{151}$$

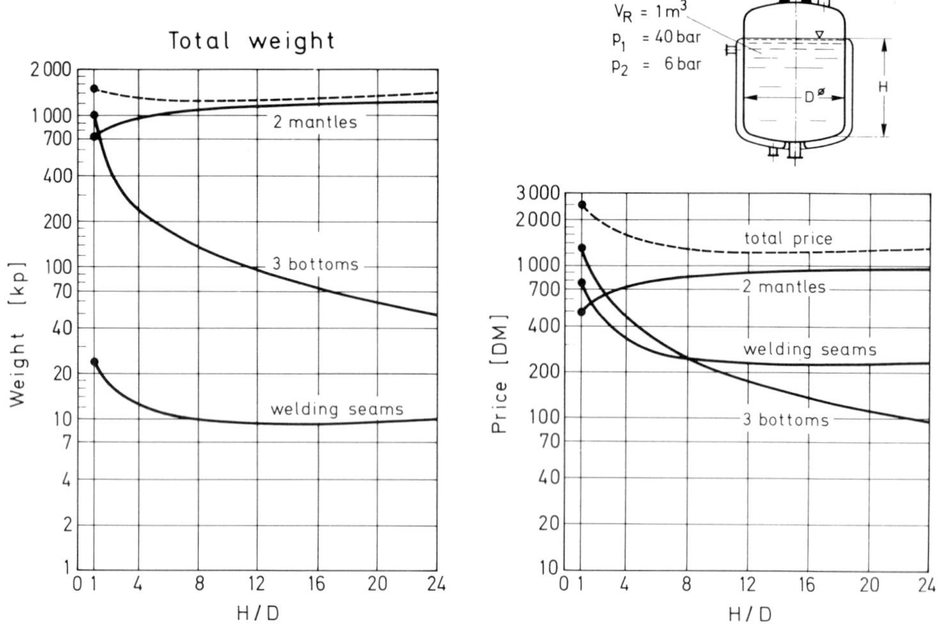

Fig. 71. Influence of the grade of slenderness *s* on weight and price of a welded pressure vessel

or for the "reference reactor"

$$\dot{q}_0 = \dot{q}_{RA} + \dot{q}_{M_0} \tag{152}$$

i.e., with increasing a' (150) the relative total costs $\frac{p_c}{p_{c_0}}$ can be reduced at $\frac{s}{s_0}$ = const. more. Furthermore for all a'-values the relative total costs $\frac{p_c}{p_{c_0}}$ reach minimum values in the range of

$$s_{opt} \approx (3 - 3.5)\, s_0 \approx 15 - 18$$

where they can be reduced to 50 or even 40% of those of the *"reference JLR"* with s_0 = 5. This economic optimization of the *JLR* thus leads to slim reaction tower types, similar to distillation columns.

Following this, brings a further lowering of costs, which we did not mention up to now, namely the construction costs, as Fig. 71 indicates[46]. According to this the construction costs can be approximately halved by changing from $s \approx 1$ (conventional *STR*) to $s \approx 16$ (slim tower type *LR*).

The minimum of construction costs (Fig. 71) is in amazingly good agreement with the minimum of capital and energy costs of Fig. 70. Although these latter economic considerations are not yet quantitatively exact, they do show however the tendency that *LR* with higher grades of slenderness $s = \frac{H}{D}$ lead to lower construction, capital, and operation costs.

9 Examples of Actual Research in the Development of LR

9.1 Optimal O_2-Conversion

Let us finally discuss only roughly and in a qualitative manner just one actual theoretical problem we are studying, namely the concept to determine the optimal O_2-conversion $\overline{u}_{O_2}$ of the air throughput in aerobic reactions within the *L*-phase of *G–L*-systems, for instance *SCP*-systems, using the prediscussed results of our investigations with the *sulfite system.* This problem is worked out in detail within a thesis and discussed by *R. Seipenbusch* in this book.

The specific O_2-input $\dot{m}_{O_2} = \frac{\dot{M}_{O_2}}{V_L}$ requires a specific *G–L*-interface $a_L = \frac{A}{V_L}$ which is according to (112) $a_L = \frac{A}{V_L} = \frac{\dot{m}_{O_2}}{k_L\,\overline{\Delta c}}$ with $\overline{\Delta c} = \overline{c}_o - c_L$ (108) and in ideal tube characteristic of gas throughflow according to (113) and (106) and Fig. 54

$$\overline{c}_o = \frac{c_{o_a} - c_{o_\omega}}{ln\,\frac{c_{o_a}}{c_{o_\omega}}} = \varphi\,(He, \overline{u}_{O_2})\,.$$

Thus results

$$a_L = \frac{\dot{m}_{O_2}}{k_L}\,f\,(\mathrm{He}, \overline{\mathrm{u}}_{O_2})\,. \tag{153}$$

On the other hand a *JLR* can produce a specific *G–L*-interface a_L as indicated in Fig. 56 for a_R in the range of $\frac{P_L}{V_L} \approx 0.4 - 4$ kW m^{-3} and $w_G \gtrsim 4$ cm s^{-1}, when $a_L \neq f(w_G)$ corresponding to (130) and Fig. 60 considering (128) and (63)

$$a_L \approx C_1' \left(\frac{P_L}{V_L}\right)^{2/3}\,. \tag{154}$$

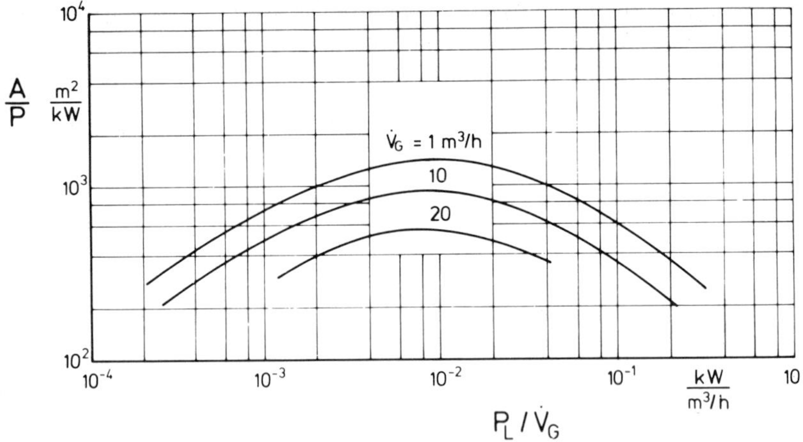

Fig. 72. Optima of $\frac{A}{P}$ as a function of $\frac{P_L}{\dot{V}_G}$ for various gas flow rates $\dot{V}_G$

Our experimental data[27, 28] led, as shown in Fig. 72, which we already published in[21], to a significant maximum of $\frac{A}{P}$ as a function of $\frac{P_L}{\dot{V}_G}$, and this independent of $\dot{V}_G$ within our experimental range around

$$\left(\frac{P_L}{\dot{V}_G}\right)_{opt} = \left(\frac{\frac{P_L}{V_L}}{\frac{\dot{V}_G}{V_L}}\right)_{opt} = 0.01 \, . \tag{155}$$

The mass balance of O_2 for the air throughput results in the O_2-conversion corresponding to (115)

$$\overline{u}_{O_2} = \frac{\dot{M}_{O_2}}{\dot{V}_G \, c_{G_a}} = \frac{\dot{m}_{O_2}}{\frac{\dot{V}_G}{V_L} c_{G_a}} \tag{156}$$

or

$$\frac{\dot{V}_G}{V_L} = \frac{\dot{m}_{O_2}}{c_{G_a} \, \overline{u}_{O_2}} \, . \tag{157}$$

Putting (157) into (155) leads to

$$\left(\frac{P_L}{V_L}\right)_{opt} \approx 0.01 \, \frac{\dot{m}_{O_2}}{c_{G_a} \, \overline{u}_{O_2}} \tag{158}$$

and (158) into (154) to

$$a_{L_{opt}} = C_2' \left(\frac{\dot{m}_{O_2}}{c_{G_a}\, \overline{u}_{O_2}} \right)^{2/3} . \tag{159}$$

Equation (159) with (153) yields

$$\frac{\dot{m}_{O_2}}{k_L} f(He, \overline{u}_{O_2}) = C_2' \left(\frac{\dot{m}_{O_2}}{c_{G_a}\, \overline{u}_{O_2}} \right)^{2/3} . \tag{160}$$

In (160) the following values are known or predetermined:

$$\dot{m}_{O_2};\ k_L;\ He;\ C_2';\ c_{G_a} .$$

Thus it can be solved to give the optimal O_2-conversion $(\overline{u}_{O_2})_{opt}$ of the air throughput, which requires the minimal power input $\frac{P_L}{V_L}$ (158) to realize the optimal specific *G–L*-interface $a_{L_{opt}}$.

9.2 Optimal Design, Operation, and Combinations of LR

In addition to the prediscussed actual *theoretical* considerations let us just briefly indicate an example of *constructive-operational* developments, which just keep us busy. It was explained in Sect. 6 that the specific oxygen input $\dot{m}_{O_2}$ (112) is proportional to the driving concentration difference $\overline{\Delta c} = \overline{c}_o - c_L$ (108) for constant values of k_L (110) and a_L (111), as illustrated in Fig. 53. Let us consider again the overall mean O_2-concentration $\overline{c}_o$ in the *L*-phase just at the *G–L*-interface, while we presume in good approximation $c_L = const.$ because of the high turbulence within the *L*-phase. At throughput of air without any mixing between *G*-bubbles (segregated *G*-system corresponding to *ideal tube flow*) we find considerable larger $\overline{c}_o$-values (113) than in the case of intense mixing, especially backmixing, of the *G*-phase where in the case of an ideal *STR* $\overline{c}_o = c_{o_\omega}$ (114), as Fig. 54 shows.

The usual types of *LR* are just characterized by the backmixing effect due to recirculation of all phases, improving the degree of mixing (Sect. 5.2), which is desired concerning distribution of substrate and oxygen in the *L*-phase and equalization of temperature in the whole reaction volume as well as redispersion of gas bubbles, paraffine drops and cell agglomerates. In general with the *L*-phase also the *G*-phase is recirculated as in Fig. 7. This is also profitable concerning the redispersion of coalesced gas bubbles in the shear fields of *L*-jets or propellers, thus increasing a_L (111, 112). But on the other hand unfortunately the recirculation of *G*-bubbles with low O_2-concentration depresses $\overline{c}_o$, at intense circulation flow approximately to $\overline{c}_o \approx c_{o_\omega}$ (114), thus reducing $\overline{\Delta c}$ (108), and $\overline{\dot{m}}_{O_2}$ (112).

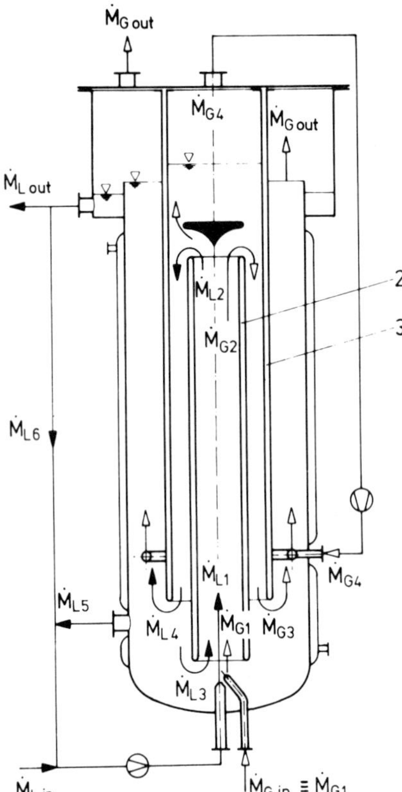

Fig. 73. *JLR* with recirculation of L-phase and direct throughflow of L-phase, leading to a combination of a thus modified *LR* with a bubble column (*BC*) in series, all in one apparatus

At the end of Sect. 6.3 we already mentioned an *LR* operation with *LR*-characteristic (recirculation) of the *L*-phase, whilst the *G*-phase passes the *LR* in *tube-flow*-characteristic, thus realizing the higher $\overline{c}_o$ (113). However this mode of operation only allows very small gas throughputs.

But there is another type of *LR* design and operation, which realizes recirculation of the *L*-phase and straight throughflow of the *G*-phase according to Fig. 73 for the same high *G*-throughputs as in conventional *LR*-types. In this new type[47)] the *G*-phase (open arrows) is branched off just before the recirculation in the lower *LR* part and led to an external annular space, where it bubbles up as in a *bubble column (BC)* part. In this example of Fig. 73 the partial *G*-flow $\dot{M}_{G4}$ leaving the *LR* part above is led back into the *BC* part for further O_2-conversion.

Such a combined *LR–BC*-reactor could be varied in many ways[47)], for instance by adding a further *BC* part above, as indicated in Fig. 74.

Anyhow, in this sense many combinations, e.g., of *LR*, *tube reactor*, and *bubble column* behaviour can be realized for the different phases to satisfy favourable reaction conditions in very simple ways of construction and operation.

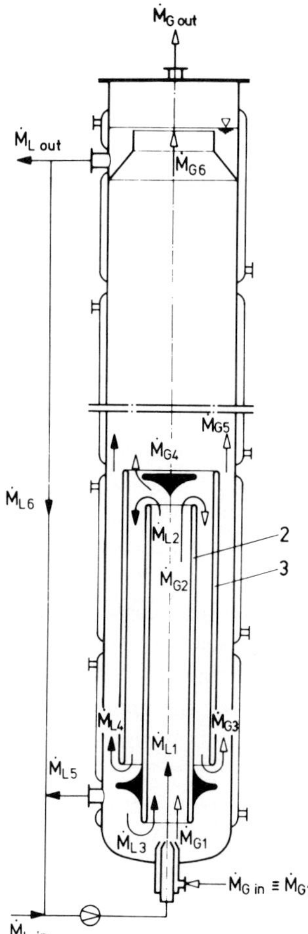

Fig. 74. Combined modified *LR* with *BC* in series as in Fig. 73 and an additional *BC* above it, all in one apparatus

10 Summary and Outlook

Let us briefly summarize the most important advances of *LR* mainly with regard to their application as bioreactors:

1. Simple slender construction with few, simple, and smooth installations; no moving parts and shaft sealings (besides *PLR*); thus low investment costs (apparatus, drive, fundaments); easy to sterilize and to keep sterile.
2. Defined low-loss flow direction in the whole reaction space, e.g., by draft tubes; defined distribution of all components according to the biochemical demands; low specific power requirement for the demanded degree of mixing mainly by appropriate design of guiding devices.
3. Very efficient primary dispersion of the *G*-phase (mainly in *JLR*) to produce large specific *G–L*-interface for the mass transfer between both phases, as well as of other

dispersed phases such as paraffin as a substrate; in addition efficient dispersion and intense mass transfer coefficient throughout the whole reactor volume in consequence of high turbulence (Re_m); low specific power requirement for those dispersing and mass transfer effects.

4. Effective additional redispersion of all disperse phases, such as coalesced gas bubbles or paraffin drops and cell agglomerates by the defined circulation flow, which is characteristic for the *LR* and forces repeated passages through the intense shear fields around the *L*-jet or the propeller.
5. Negligible cell damage by the circulation drive mainly in *ALR* and *JLR* with their relative smooth impulse transfer.
6. Large specific cooling areas by the slim shape of *LR* and the simple opportunity, to use guiding devices for additional cooling; therefore it is possible to realize very large reactors without external cooling circuits.
7. High heat (as mass) transfer coefficients in consequence of high velocities and turbulence at contact surfaces and interfaces, and that with low power requirement.
8. Defined generation of gas hold up ϵ and circulation flow (Re_m, w_m) with *JLR* by realizing a certain Re_1 at a certain P_L; on the other hand defined generation of *G–L*-interface a_L by realizing a certain P_L at a certain Re_1, both by selection of the appropriate diameter D_1 of the *L*-nozzle.
9. Self-acting intense suction of foam and flotating material by the downdraft circulation flow; the effect can be intensified, e.g., by appropriate flow guiding devices around the upper end of the draft tubes.
10. Self-acting whirling up of sedimenting materials such as cell agglomarates from the bottom of the reactor by appropriate design and flow guiding in this area and by sufficient circulation flow ($\check{R}e_m$).
11. Defined suction or separation of gas in *G–L*-systems by appropriate flow guiding devices (e.g., gas deflectors).
12. Various combinations of design and operation of *LR* itself and of its combination, e.g., with tube reactors or bubble columns in simple ways of construction and operation mainly by appropriate flow guiding. Simple adaptation to operational or process changes.

But in spite of all these advantages there have still to be solved some important problems to enable calculation and design for safe and optimal layout, construction, and operation of *large scale* production plants for the broad variety of biotechnological requirements. Some of those problems are for example:

- Optimal combination of construction elements and operation conditions for large scale plants, especially concerning local distributions of shear fields, turbulence, and "*specific*" values as $\frac{P}{V}$ or $a = \frac{A}{V}$ throughout the whole reaction space.
- Determination of local concentration distributions and individual residence time behaviour of all phases depending on fluid dynamic parameters.
- Investigation of fluid dynamics in highly viscous aerated systems.
- Transfer and application of experimental results from simple model systems to complicated biosystems and from small model apparatuses to large scale production plants.

Finally it may be stated that all types of *LR* in consequence of the relative short time of application in chemical and biochemical engineering surely justify some expectations in competition with other reactor types. But nevertheless *LR* still require considerable efforts in research and development and there is no doubt that however effective any reactor type may prove to be, there is no chance to find out the optimal universal bioreactor (or chemical reactor)[48]. The extreme variety of (bio-)chemical systems and processes will always require different reactor types. The *LR* may be a favourable one for some fields, and other types will be more appropriate for other fields or application.

11 Nomenclature

a	$m^2\ s^{-1}$	heat conductivity number
a	m	distance of draft tube from jet nozzle
a'		ratio of reaction heat + heat capacity of feed and mixing power of reference reactor
a_C	m^{-1}	specific cooling area
a_L	m^{-1}	specific G–L-interfacial area ref. to liquid volume
a_R	m^{-1}	specific G–L-interfacial area ref. to reactor volume
A	m^2	G–L-interfacial area
A_a	m^2	cross section of annulus
A_o	m; cm	distance of draft tube from surface
A_u	m; cm	distance of draft tube from bottom
A_B	m^2	cooling area of bottom
A_C	m^2	cooling area
A_D	mm	distance of jet nozzle from draft tube
A_E	m^2	cooling area of draft tube
A_G	m^2	outlet area of gas distributor
A_M	m^2	cooling area of mantle
b	m^2	relative heat transfer coefficient
c	kmol m^{-3}; kg m^{-3}	concentration
$\hat{c}$	kmol m^{-3}; kg m^{-3}	concentration of tracer at impuls maximum
c_a		solid particle concentration
c_c		ratio of energy and capital costs
c_{in}	kmol m^{-3}	tracer concentration in the inlet
c_o	kmol m^{-3}; kg m^{-3}	liquid side oxygen concentration at G–L-interface
$\bar{c}_o$	kg m^{-3}; kmol m^{-3}	log. mean concentration allover G–L-interface
c_{o_a}	kg m^{-3}	G–L-interface oxygen concentration at air inlet
c_{o_ω}	kg m^{-3}	G–L-interface oxygen concentration at air outlet
c_r		relative concentration
$\hat{c}_r$		relative concentration at impulse maximum
c_G	kg m^{-3}	oxygen concentration in G-phase
c_L	kg m^{-3}	oxygen concentration in L-phase
c_{G_a}	kg m^{-3}	oxygen concentration in G-phase at air inlet

c_{R_o}		volume-concentration at uniform distribution throughout reactor volume
c_{Sub}	kmol m^{-3}	substrate concentration
c_T	kg m^{-3}	tracer concentration in the outlet
c_1	kmol m^{-3}	concentration of reactant 1
c_{1t}	kmol m^{-3}	concentration of reactant 1 at time t
c_∞	kg m^{-3}	mean tracer concentration for total tracer mass distributed in reactor
Δc	kg m^{-3}	driving concentration gradient
$\overline{\Delta c}$	kg m^{-3}	mean driving concentration gradient
C		constant
C_1		constant
C_1'		constant
C_2		constant
C_2'		constant
C_3		constant
C_4		constant
C_5		constant
$\bar{d}_B$	mm	mean bubble diameter
d_E		diameter ratio
D	m; mm	reactor diameter
D_1	mm	diameter of liquid jet nozzle
D_{eff}	m^2 s^{-1}	effective longitudinal diffusivity
D_E	m; mm	mean diameter of draft tube
D_{Ei}	m; mm	internal diameter of draft tube
D_L	m^2 s^{-1}	diffusivity of gas molecules in liquid
D_P	m	diameter of propeller
F_{ALR}	N	air-lift driving force
F_I	N	resistance force
F_{II}	N	resistance force
g	m s^{-2}	gravity constant
h		inhomogenity
$\hat{h}$		inhomogenity ref. to impulse maximum
H	m	reactor height
H_P	m	pressure head of propeller
H_B	m; cm	aeration height
H_G	m	settling height
$H(\tau)$		distribution function
He		Henry coefficient
i		number of completed circulations
i_M		number of circulations to obtain required degree of mixing
i_U		number of circulations within time t
$\dot{I}_I$	N	momentum flow entering the draft tube
$\dot{I}_{II}$	N	momentum flow leaving the draft tube
k		constant
k	kW m^{-2} K^{-1}	overall heat transfer coefficient
k'	s^{-1}	first order reaction rate constant

k_L	$m\ s^{-1}$; $m\ h^{-1}$	mass transfer coefficient
k_1		correction factor for liquid jet effect
k_2	$m^3\ kmol^{-1}\ s^{-1}$	2nd order reaction rate constant
k_s		correction factor for height to diameter ratio
K	$Pa\ s^n$	consistency factor
l_E		length to diameter ratio
L	m	length
L_E	m	length of draft tube
L_U	m	length of one circulation
m		capital costs' exponent
$\dot{m}_{O_2}$	$kg\ m^{-3}\ h^{-1}$	specific oxygen flow
$\dot{m}_{Sub}$	$kg\ m^{-3}\ h^{-1}$	specific substrate mass flow
$\dot{M}_1$	$kg\ s^{-1}$	mass through flow
$\dot{M}_2$	$kg\ s^{-1}$	circulation mass flow
$\dot{M}_3$	$kg\ s^{-1}$	total mass flow
M_0	kg	tracer mass
M'	kg	amount of tracer mass having left the reactor
M_{O_2}	kg	oxygen mass
$\dot{M}_{O_2}$	$kg\ s^{-1}$	oxygen mass flow
M_R	kg	reaction mass
n		flow index
n_{eq}		equivalent number of consecutive equal-volume ideal *STR*s
n_L		rotation number in *L–S*-system
n_P	s^{-1}; rpm	propeller rotation speed
$\dot{n}_{O_2}$	$kmol\ m^{-2}\ h^{-1}$	oxygen molar flow density
n_{Sus}		circulation number of suspension
n_U		circulation number
$\dot{N}_{O_2}$	$kmol\ h^{-1}$	oxygen actual molar flow
N_1	kmol	molar mass of reactant 1
p	bar	pressure
p	$kg\ m^{-3}\ h^{-1}$	specific productivity
p_1	bar	dynamic pressure
p_c	$DM\ m^{-3}$	specific total costs of reactor
p_{c_0}	$DM\ m^{-3}$	specific total costs of reference reactor
p_{L_0}	bar	excess static pressure at nozzle outlet
p_P	bar	propeller pressure
Δp_U	bar	pressure drop of circulation
P	kW	total power input
P_G	kW	aeration power
P_L	kW	liquid jet power
P_{L_0}	kW	excess static pressure power
P_P	kW	pumping power of propeller
$\dot{q}$	$kW\ m^{-3}$	specific total caloric power
$\dot{q}_0$	$kW\ m^{-3}$	specific total caloric power of reference reactor
$\dot{q}_C$	$kW\ m^{-3}$	specific caloric power removed
$\dot{q}_{M_0}$	$kW\ m^{-3}$	specific mixing power of reference reactor

$\dot{q}_M$	kW m^{-3}	specific mixing power of reactor for increased s
$\dot{q}_{RA}$	kW m^{-3}	specific reaction heat flow + heat capacity of feed flow
r_U	s^{-1}	circulation rate
s		slenderness ratio
s_0		slenderness ratio of reference reactor
s_E	m	thickness of draft tube wall
$S(\tau)$		transition function
t	s	time
$\bar{t}$	s	mean residence time
t_{eq}	s	equivalent residence time (penetration theory)
t_i	s	time for i circulation
$\bar{t}_G$	s	mean residence time of gas
t_M	s	mixing time
t_U	s	circulation time
$\bar{t}_U$	s	mean circulation time
T	K; °C	temperature
T_R	K; °C	reaction temperature
$\bar{T}_C$	K; °C	logarithmic mean coolant temperature
ΔT_m	K; °C	mean temperature difference
$\bar{u}$		mean conversion rate
$\dot{V}_1$	$m^3\ s^{-1}$; $l\ s^{-1}$	liquid volume through flow
$\dot{V}_2$	$m^3\ s^{-1}$; $l\ s^{-1}$	liquid volume circulation flow
$\dot{V}_3$	$m^3\ s^{-1}$; $l\ s^{-1}$	total liquid volume flow
V_G	m^3	gas volume in G–L-system
$\dot{V}_G$	$m^3\ s^{-1}$	gas volume through flow
V_L	m^3	liquid volume in G–L-system
$\dot{V}_L$	$m^3\ s^{-1}$; $l\ s^{-1}$	liquid volume through flow in G–L-system
$\dot{V}_P$	$m^3\ s^{-1}$	propeller volume flow rate
V_R	m^3	reactor volume
ΔV	m^3	volume element
w	m s^{-1}	velocity
w_1	m s^{-1}	liquid velocity in liquid nozzle
w_2	m s^{-1}	velocity of circulation liquid entering draft tube
w_i	m s^{-1}	liquid mean velocity in draft tube after impulse transfer in the outlet of draft tube
w_m	m s^{-1}	mean liquid velocity of circulation
w_s	m s^{-1}	settling rate
w_G	m s^{-1}; cm s^{-1}	superficial gas velocity
w_{L_a}	m s^{-1}	liquid velocity in external annulus
w_{L_m}	m s^{-1}	mean liquid velocity in heterogenous systems
X_o		dimensionless upper distance number of draft tube
X_u		dimensionless lower distance number of draft tube
x	kg m^{-3}	cell concentration
x_U		relative flow path

Greek Letters

α	kW m^{-2} K^{-1}	heat transfer coefficient
$\dot{\gamma}$	s^{-1}	shear rate
δ	m	film thickness
ϵ		gas hold up
ϵ_a		gas hold up in annulus
η	Pa s	dynamic viscosity
η_s	Pa s	appearent liquid viscosity
μ	h^{-1}	specific growth rate
ν_1	m^2 s^{-1}	viscosity of liquid entering through jet nozzle
ν_L	m^2 s^{-1}	liquid viscosity in *G–L*-system
ν_m	m^2 s^{-1}	mean liquid viscosity of total flow
ζ_a		resistance number for external annular flow
ζ_i		resistance number for draft tube flow
ζ_o		resistance number for flow round upper edges of draft tube
ζ_u		resistance number for flow round lower edges of draft tube
ζ_U		circulation resistance number for *JLR*
ζ_U^*		circulation resistance number for *LR* without liquid jet
ρ	kg m^{-3}	density
ρ_1	kg m^{-3}	density of liquid entering through jet nozzle
ρ_m	kg m^{-3}	mean liquid density of total flow
$\Delta\rho$	kg m^{-3}	difference between mean density of *G–L* system inside draft tube and external annulus, respectively
τ	Pa	shearing stress
τ		relative residence time
τ_i		relative time for *i* circulation
τ_M		relative mixing time
τ_U		relative time related on circulation time

Dimensionless Numbers

Bg	geometric characteristic number
Bg'	geometric characteristic number
Bo	Bodenstein number
Eu	Euler number
Fr	Froude number
Ne	Newton number
N_V	volume flow number
Pe	Peclet number
Pr	Prandtl number
Re_1	Reynolds number for liquid nozzle
Re_m	Reynolds number ref. to circulation
Re_P	Reynolds number ref. to propeller
Re_P^+	modified Reynolds number ref. to propeller
Sc	Schmidt number

Subscripts and Symbols

a	annular
c	costs

B	bottom
B	bubble
C	cooling
eff	effective
eq	equivalent
E	draft tube
G	gas
i	inside
i	interface
i	after *i* circulations
l	laminar
L	liquid
m	mean
M	mixing
o	upper
o	interface on liquid side
opt	optimal
O_2	oxygen
P	propeller
Pen	penetration theory
r	relative
real	real
R	reactor
s	apparent
s	sinking
Sub	substrat
Sus	suspension
S	solid
t	turbulent
u	lower
U	circulation
α	at air inlet
ω	at air outlet
0	status as $t = 0$
1	liquid nozzle
2	circulation
3	total flow
∞	infinite
$\wedge$	maximum
$\vee$	minimum
$-$	mean

Abbreviations

BC	bubble column
CMC	carboxymethylcellulose
JLR	jet loop reactor
LR	loop reactor
ALR	air-lift-loop reactor
PLR	propeller loop reactor
STR	stirred tank reactor
SCP	single cell protein

12 References

1. Blenke, H., Schlingmann, M., Sittig, W.: C.P.C.I.A-Progrès et développement de la technologie alimentaire. Paris 1977
2. Blenke, H.: VDI-Berichte Nr. 277, 127 (1977)
3. Blenke, H., Bohner, K., Schuster, S.: Chem.-Ing.-Techn. *37*, 289 (1965)
4. Wang, D.I.C., Humphrey, A.E.: Chem. Eng. *26*, 108 (1969)
5. Slater, L.E.: Food Eng., 68, Juli 1974
6. Littlehailes, J.D.: 1. Symp. Mikrobielle Proteingewinnung, Stöckheim 1975, p. 43
7. Blenke, H.: Chem.-Ing.-Techn. *39*, 109 (1967)
8. Blenke, H., Bohner, K., Pfeiffer, W.: Chem.-Ing.-Techn. *43*, 10 (1971)
9. Blenke, H., Bohner, K., Hirner, W.: Verfahrenstechnik *3*, 444 (1969)
10. Blenke, H., Bohner, K., Vollmerhaus, E.: Chem.-Ing.-Techn. *35*, 201 (1963)
11. Stein, W.: Chem.-Ing.-Techn. *40*, 829 (1968)
12. Blenke, H., Seipenbusch, R.: Verfahren und Vorrichtung zur aeroben Fermentation. P. 24 36 793.9
13. Marquart, R., Blenke, H.: Verfahrenstechnik *12*, 721 (1978)
14. Marquart, R.: Verfahrenstechnik *13*, 527 (1979)
15. Zlokarnik, M.: Rührtechnik. In: Ullmanns Enzyklopädie der Technischen Chemie. 4. Aufl. Bd. 2, p. 259. Verlag Chemie Weinheim 1972
16. Metzner, A.B., Otto, R.E.: AICHE Journal *3*, 3 (1957)
17. Pfeiffer, W., Blenke, H., Muschelknautz, E.: Verfahrenstechnik *11*, 95 (1977)
18. Bohner, K., Blenke, H.: Verfahrenstechnik *6*, 50 (1972)
19. Ilg, A.: Gasgehalt, Phasengrenzfläche und Umwälzung im Mammutschlaufenreaktor. Diplomarbeit Universität Stuttgart 1977
   Raff, M.: Fluiddynamische Untersuchungen am Schlaufenreaktor SR 630. Diplomarbeit Universität Stuttgart 1976
   Reule, W.: Messung des flüssigkeitsseitigen Stoffübergangskoeffizienten für die Sauerstoff-Übertragung aus Luft in eine Bakteriensuspension. Diplomarbeit Universität Stuttgart 1976
   Schumm, W.: Messung der örtlichen Sauerstoffkonzentration in der Gas- und Flüssigphase im Schlaufenreaktor am System Sulfit-Luft. Diplomarbeit Universität Stuttgart 1976
   Vogel, H.: Untersuchungen neuartiger Varianten des Schlaufenreaktors. Diplomarbeit Universität Stuttgart 1978
20. Seipenbusch, R., Blenke, H., Birkenstaedt, J., Schindler, F.: 5. Int. Ferment. Symp. Berlin 1976
21. Seipenbusch, R., Birkenstaedt, J., Schindler, F.: 1. Symp. Mikrobielle Proteingewinnung. Stöckheim 1975, p. 59
22. De Waal, K.J.A., Okeson, J.C.: Chem. Eng. Sci. *21*, 559 (1966)
23. Linek, V., Mayrhoferova, J.: Chem. Eng. Sci. *25*, 787 (1970)
24. Wesselingh, J.A., Van't Hoog, A.C.: Trans. Instn. Chem. Eng. *48*, T 69–T 74 (1970)
25. Bell, H.G., Gallo, M.: Proc. Biochem., p. 33, April (1971)
26. Zlokarnik, M.: Europ. Fed. Biotech, Interlaken 1978
27. Blenke, H., Hirner, W.: VDI-Berichte Nr. 218, 549 (1974)
28. Hirner, W., Blenke, H.: Verfahrenstechnik *11*, 297 (1977)
29. Pawlowsky, J.: Chem.-Ing.-Techn. *34*, 628 (1962)
30. Lehnert, J.: Verfahrenstechnik *6*, 58 (1972)
31. Lehnert, J.: Berechnung von Mischvorgängen in schlanken Schlaufenapparaten. Diss. Universität Stuttgart 1972
32. Levenspiel, O., Smith, W.K.: Chem. Eng. Sci. *6*, 227 (1957)
33. Schoenemann, K.: Dechema-Monographien *21*, 203 (1952)
34. Higbie, R.: Trans. Am. Inst. Chem. Engrs. *31*, 365 (1935)
35. Danckwerts, P.V.: Ind. Eng. Chem. *43*, 1460 (1951)
36. Astarita, G.: Mass transfer with chemical reaction. Amsterdam, London, New York: Elsevier 1967

37. Lenmann, J., Oels, U., Schügerl, K.: 1. Symp. Mikrobielle Proteingew. Stöckheim 1975, p. 133
38. Reith, T.: Physical aspects of bubble dispersions in liquids. Ph. D. Thesis, T.H. Delft, 1968
39. Linek, V., Turdik, J.: Biotech. Bioeng. *13*, 353 (1971)
40. Danckwerts, P.V.: Gas-liquid reactions. New York: Mc Graw-Hill 1970
41. Linek, V., Mayrhoferova, J.: Chem. Eng. Sci. *24*, 481 (1969)
42. Blenke, H.: Int. Symp. on Mixing. C 7÷1–C 7÷55. Mons 1978
43. Raible, G.: Wärmeübergang im Schlaufenreaktor für das System Gas/Flüssigkeit. Diss. Universität Stuttgart 1976
44. Prinzing, P.: Fortschr.-Ber. VDI-Z. *3*, 29 (1970)
45. Blenke, H., Prinzing, P.: Chem.-Ing.-Techn. *41*, 233 (1969)
46. Blenke, H., Bohner, K.: VTG Ausschuß "Prozeß- und Anlagentechnik". Wiesbaden 1972
47. Blenke, H.: Verfahren und Vorrichtung zur Durchführung (bio-)chemischer Reaktionen und verfahrenstechnischer Grundoperationen in fluiden Systemen. Patentanmeldung 1978, P 28 47 443.3
48. Blenke, H., Dengler, W., Hinger, K.-J., Hirner, W., Ipfelkofer, R., Lehnert, J., Natusch, H.-J., Neukirchen, B., Pfeiffer, W., Prinzing, P., Raible, G., Stein, W.: Chemiereaktortechnik. Fortschritte der Verfahrenstechnik, Bd. 10, p. 527 (1970/71)